Women in the Technology and Innovation Ecosystem

In an increasingly interconnected world, the movement of people across borders has become a common and dynamic part of our global society. This book explores the lives of women expatriates in America, particularly in the technology and innovation industry. It highlights their resilience, adaptability, and pursuit of dreams in unfamiliar lands, while also exploring themes of identity, belonging, and cultural negotiation.

Women in the Technology and Innovation Ecosystem: From Movers to Makers in America explores the transformative journey of women relocating to America, focusing on entrepreneurship, leadership, research, education, technology, and innovation. It provides valuable data and case studies, enhancing policies and hiring programs for women entrepreneurs and offering strategies for cultural adaptation and integration in foreign business environments.

With the increasing number of women expatriates in America, who are making significant economic and financial contributions, this book serves as a valuable resource for aspiring female expatriates looking to succeed as professionals and entrepreneurs in the United States.

Women in Industry 4.0: Shaping World Foundations in Engineering, Business, and Sustainability

Series Editors: Tilottama Singh, Richa Goel, and Kyla L. Tennin

According to a World Economic Forum report, only 22% of AI professionals globally are female. This series will bring stories straight from the heart of these innovative women who have established themselves, to inspire millions and create a wave of impact. This will be a series of books and authors that will focus on sustainability through technology and entrepreneurship in Industry 4.0. The series will act as a global platform for bringing together organizations and women sharing their life cases, research, and grounded theory which will be an interesting read for engineers, businesses, government, researchers, and students working on startups and contributing to gender diversity.

Each book in the series is dedicated to illuminating women's research and achievements in key, targeted areas of the contemporary tech world and raising awareness of the pivotal work women are undertaking in areas of importance to our global community toward sustainable development. It will be a step in bringing forth the reality of what overshadows women in workplaces and the strategic attempts of organizations to pave the path for women's representation in companies and fields of engineering.

Books that focus on women in engineering and technology exploring and describing how women have greatly contributed to the world at large through these fields and their many subsections is a focal point of this series.

If you are interested in writing or editing a book for the series or would like more information, please contact Cindy Carelli, cindy.carelli@taylorandfrancis.com.

African Women Trailblazers in STEM
Pioneering Technological and Economic Advancement
Edited by Uzo Mkparu, Richa Goel, Tilottama Singh, and Jade Mkparu

Women, Tech, and Enterprise
A Path to Sustainable Equality in South Asia
Edited by Kavita Indapurkar, Neeru Sidana, Richa Goel, and Tilottama Singh

Tech Startups Driving Innovation
Empowering Women in Bangladesh
Edited by Quazi Tafsirul Islam, Richa Goel, and Tilottama Singh

Diversity in Women's Entrepreneurship
Global Sustainability in African Countries
Edited by Bola Adijatu Yaya, Elijah U. Felix, Tilottama Singh, and Richa Goel

Women in the Technology and Innovation Ecosystem
From Movers to Makers in America
Edited by Manleenjot Kaur, Melissa Hyunh, and Tilottama Singh

Women in the Technology and Innovation Ecosystem

From Movers to Makers in America

Edited by
Manleenjot Kaur
Melissa Hyunh
Tilottama Singh

CRC Press
Taylor & Francis Group
Boca Raton London New York

CRC Press is an imprint of the
Taylor & Francis Group, an **informa** business

Designed cover image: Shutterstock

First edition published 2026
by CRC Press
2385 NW Executive Center Drive, Suite 320, Boca Raton FL 33431

and by CRC Press
4 Park Square, Milton Park, Abingdon, Oxon, OX14 4RN

CRC Press is an imprint of Taylor & Francis Group, LLC

ISBN: 978-1-032-99481-9 (hbk)
ISBN: 978-1-032-98539-8 (pbk)
ISBN: 978-1-003-60440-2 (ebk)

DOI: 10.1201/9781003604402

Typeset in Times
by SPi Technologies India Pvt Ltd (Straive)

Contents

Preface

When we first began shaping this book, we weren't merely compiling stories—we were responding to a silence.

Across boardrooms, innovation hubs, STEM labs, start-up pitch decks, and policy tables in the United States, women from around the world are making remarkable contributions. Yet, the stories of women expatriates—those who left behind home countries, families, and familiar systems—often remain invisible or simplified. Too often, they are viewed through the narrow lenses of immigration status or cultural adjustment, rather than as architects of innovation, strategy, and transformation. This book is an attempt to change that.

Women in the Technology and Innovation Ecosystem: From Movers to Makers in America grew from a shared question: How do women who cross borders become builders of the future in one of the most demanding innovation ecosystems in the world?

As editors, we bring to this work a diversity of disciplinary lenses—ranging from entrepreneurship, gender studies, and migration to technology, organizational behavior, and policy. What united us was a shared commitment to bringing marginalized voices to the center, not only as subjects of inquiry but as agents of change.

The title "From Movers to Makers" reflects a powerful transition. Many of the women profiled here began their American journeys as spouses, students, or job-seekers—roles too often defined by dependency or uncertainty. Over time, through resilience, learning, and determination, they became makers: of companies, code, communities, careers, and change. Their stories challenge dominant narratives about who leads in tech, who innovates, and who belongs.

This book blends personal narratives, thematic insights, and evidence-based research to create a textured portrait of expatriate women in American innovation. Each chapter invites you to see these women not as "exceptions," but as representatives of a growing yet under-acknowledged force shaping global futures.

We hope this book will resonate with multiple audiences—researchers, educators, policymakers, employers, and especially other women undergoing paths across geographies and industries. Our aim is to inform, inspire, and ignite conversation around inclusion, migration, gender, and the future of innovation.

This work is also a tribute—to the women who trusted us with their journeys, to the communities that supported them, and to the ones still in transition, still navigating, still becoming.

We thank you for holding this book in your hands, and we invite you to listen deeply to the voices within.

With warmth and solidarity,
The Editors

Foreword

This book is a celebration of that courage. This book gives a peculiar and enlightening perspective on the experiences of women who go abroad and cross boundaries—physical, cultural, social, and professional—to establish themselves across the US technology and innovation ecosystem.

In this book, careers are not merely exemplified but the resilience, challenging conventions, redrafting stories, facing odds, and reimagining leadership are.

The stories shared in the book do not talk about easy victory. They narrate about grit, navigating systems that were not once designed with women in mind, finding community and strength in unexpected places, and redefining success on their own terms. They reveal the layers of identity that women carry with them as immigrants, professionals, creators, and as visionaries that enrich the global innovation ecosystem in ways we are only beginning to understand.

In a world where migration is often politicized and women's voices are too often marginalized, this book reminds us that talent knows no geography, and innovation thrives on diversity. It challenges us to examine not only how far these women have come but also how far we, as a global community, must go to build systems that nurture rather than hinder such potential.

As you turn these pages, you will meet women who are more than movers; they are makers. They are the architects of solutions, builders of businesses, and catalysts of change. Being an expatriate myself, it can be analogously said that when women move, they carry far more than their own aspirations—they move entire communities forward, shifting the boundaries of what is possible.

May this book not only inspire you but also invite you to act—to create spaces where more women can innovate freely, lead boldly, and dream expansively. When we invest in the courage and creativity of women who traverse international borders, we not only pay tribute to their experiences but also forge a more equitable, diverse, and highly inventive future.

Sarabjot Thapar
Respiratory Therapist
Medical City, Plano
USA

Acknowledgment

We start with profound and humble thankfulness to God, the Universal Creator, whose grace led this journey from the earliest of beginnings. In times of peace and in times of uncertainty, it was God's grace that provided the sustenance to carry on. This book is a humble contribution in the service of a much greater cause—raising up the unheard, the unseen, and the brave.

To the Series Editors, we are grateful for having opened the doors to this urgent and much-needed dialogue. Your commitment to this project and your insightful comments along the way were instrumental in developing the book's vision and tone.

We thank each other, as co-editors, for the joint effort, respect, and teamwork that drove this work. We each contributed different skills, but we were united in one: presenting the women expatriates' technology and innovation experiences honestly, with care, and integrity.

To the remarkable women expatriates who gave us their stories—thank you. Your stories are the center of this book. You trusted us with your lived stories, and we hope we have been faithful to the realities you so courageously uncovered. Your voices will resonate far beyond these pages.

To the researchers, authors, and scholarly reviewers—your feedback, information, and observations contributed depth and richness to this transdisciplinary endeavor. We are very grateful for the intellectual community that came together behind this work and lent it form with such dedication.

Finally, to our families, who waited patiently through long hours of writing, cancelled weekends, and emotional swings—thank you. Your love grounded us. To our children, particularly the youngest of our group, who grew up with this book—you are the future we write for. May you inherit a world that welcomes all your potential.

With gratitude and hope,
The Editors

Editors

Manleenjot Kaur is an Assistant Professor in the School of Business and Management, DES Pune University, Pune. She received her PhD in International Business from Manav Rachna International University, Faridabad. She has also presented studies on women entrepreneurs, expatriates, and expat-preneurs in national and international conferences and various international platforms. She has embraced diverse management roles in various organizations, but presently, she takes pride in imparting knowledge and guiding bright minds in International Business, Entrepreneurship Development, and Strategic Management. Her teaching experience goes beyond the classroom, as she has actively organized engaging events such as the Entrepreneurship Conclave, webinars, Faculty Development Programs, and seminars. On the entrepreneurial front, running a successful franchise has allowed her to wear the hat of a skilled businesswoman, involved in every aspect from market research to digital marketing strategies. She has published various research articles in Scopus, Web of Science, and international journals, and her primary focus of research centers around women entrepreneurship, strategy, international business, innovation, and cross-cultural studies.

Melissa Hyunh is a Respiratory Therapist in Texas, USA. She studied respiratory therapy at Vanier College. Melissa has around 10 years of enriched experience in the healthcare industry. Melissa is inclined toward the research on women leadership, strategy, healthcare management, organizational dynamics, and women empowerment. She focuses on the diverse leadership styles, examining barriers to advancement and analyzing successful strategies for women leaders which can foster more inclusive and equitable workplaces. She intends to shed light on the broader societal implications of gender diversity in leadership roles, influencing policies and practices that promote equality and empower women in leadership positions.

Tilottama Singh is a certified HR analyst from IIM and a proficient academic, researcher, and trainer with more than 14 years of experience in the field of Human Resources and Work Dynamics. She is currently working as a Professor and Head of Department in the School of Management, Uttaranchal Institute of Management, Uttaranchal University, Dehradun. Her areas of expertise are Human Resources and Strategy, and her teaching concentration includes Human Resources, Strategy, and Law. She started her career in the hotel industry with Leela Kempinski in the corporate head office in Human Resources and Training. Later, she joined the education sector with the topmost universities like IMS Unison Group and Amity University, the leading university in Pan Asia. She has obtained her Master's in Human Resources Management with Distinction. Prior to that, she graduated in economics with honors and law with a distinction, with an added vocational course in mass communication and analytic certification from IIM. Having an enriched research portfolio with Scopus and international refereed journals, book chapters, and conference presentations, she has been awarded as Best Presentation Award in international conferences like IEOM, MIDAS, and many more. She also serves as a member of AIMA and acts to lead and liaise between the student community and the industry delegates, with a keen interest in training and building sustainable models for business and society.

Contributors

Amy Amann
HR Consultant and USA repatriate
Minneapolis, Minnesota, USA

Sonam Bhaduriya
Assistant Professor
Swami Rama Himalayan University
Dehradun, India

Sheela Bijlwan
Assistant Professor
Swami Rama Himalayan University
Dehradun, India

Veerle Beelen Da Costa David
Certified Travel Coach and Life Coach, Founder of Unpackbyveer
Sydney, Australia

Izabela Jastrzebowska
Certified Psychologist and Jungian Coach
Aha Moment Coaching
San Francisco Bay, California, USA

Surbhi Kapur
Professor
Manav Rachna International Institute of Research and Studies (MRIIRS)
Faridabad, India

Harleen Kaur
Associate Professor
University School of Business
Chandigarh University
Gharuan, Punjab, India

Sushil Kumar
Global Professor of Practice
Golden Gate University
San Francisco, California, USA

Sandra Morno
Tax Strategist for U.S. Expats
SSBM Consulting
Miami, Florida, USA

Giulio Palmas
PhD Student
University of Cagliari
Cagliari, Italy

Praveen Kumar Pandey
Assistant Professor
New Delhi Institute of Management
New Delhi, India

Mohd Asif Shah
Associate Professor and Dean
Kardan University
Parwan Du, Kabul, Afghanistan

Anuja Shukla
Associate Professor
Jaipuria Institute of Management
Noida, India

Rashmi Singh
Associate Professor
Jaipuria Institute of Management
Lucknow, India

Rosy Singh
Associate Professor
University School of Business
Chandigarh University
Gharuan Punjab, India

Jyoti Suri
Assistant Professor
Lingaya's Vidyapeeth
Faridabad, India

Introduction

I am not free while any woman is unfree, even when her shackles are very different from my own.

—Audre Lorde

Every woman who crosses borders in search of opportunity carries more than just luggage—she carries dreams, doubts, resilience, and often, a quiet determination to begin again.

This is their story!

Women who uprooted from their home country to construct lives in the United States—not as experts, but as creators within the technology and innovation space. Some came with degrees in hand and jobs lined up. Others came for partners or bet on themselves. Most began on the periphery—new accents, new systems, unrecognizable codes (tech and cultural). But they didn't stop there.

They innovated.

With late-night coding, start-up ventures, research insights, and day-to-day acts of bravery, these women started to chip away at space for themselves. They learned to decipher not only languages but also mindsets. They bucked the double challenge of being both "foreign" and "female" in markets still figuring out how to most fully welcome either.

And what came forth instead was not simply survival—it was leadership. Innovation. Revolution.

This book is a tribute to those journeys—from movers to makers. It's built on real stories, reflections, and research that highlight not only the barriers these women face but also the incredible value they bring. You'll meet engineers, scientists, founders, designers, educators, and policy shapers. Each with a different path, but all bound by a common thread: the courage to create.

Telling these stories, this book also poses a broader question: Are we creating a world that cannot accommodate such talent to flourish? And what might we unleash—technologically, socially, humanly—if we actually did create?

Whether you are a teacher, a policymaker, an employer, a researcher, or just a fellow traveler on your own path, I hope that these pages provide more than understanding. I hope they provide inspiration.

Because when women move, they move more than geographies. They move possibilities.

1 Pre-Departure Planning *"Wherever You Go, You Take Yourself"*

Veerle Beelen Da Costa David

1.1 INTRODUCTION – THE ONE THING WE FORGET WHEN MOVING ABROAD

"If there's one thing that you could advise people who move abroad not to forget, what would it be?"

Before writing this chapter, this is the question I was asked. I didn't have to think long about the answer, because it's something that's crystal clear to me and something that I've seen people forget again and again, including myself during my first move abroad.

Here's what most people forget when they move abroad:

Themselves.

You forget yourself.

You get so consumed by all the logistics of your move that the most important asset that you'll bring along – *you* – is completely ignored.

You're busy arranging your visa and getting stressed because your immigration attorney isn't responding fast enough, or you're confused by the documents you need to provide. Your high school diploma? How are you supposed to know where that is? You haven't seen it in over 20 years!

And that's just the beginning. You're also figuring out your budget and your finances. Do you need a different credit card? How will you transfer money without losing too much on conversion costs and fees? You're ending your lease or selling your place, notifying the authorities that you're leaving, and endlessly scrolling real estate listings in your new destination, hoping to find the perfect place to live.

Then there's your stuff. Do you sell it? Donate it? Give it to friends or family? Throw it away? Ship it? And once you start packing, you realize just how much you've accumulated things you forgot you even owned. Some of it is sentimental and brings up memories. It makes you pause. But still … you haven't touched it in years. So, do you keep it or let it go?

DOI: 10.1201/9781003604402-1

And on top of all of that, there's the emotional weight of telling your loved ones you're leaving. Will they be supportive? Will they understand? Or will they overwhelm you with questions and worries? Projecting their fears onto you, as if you hadn't already considered those risks yourself?

Eventually, moving day gets closer and closer. You need to book your flight. How much luggage can you bring? When do you hand over the keys of your old apartment? Where do you stay in the final days before departure? Will the relocation company show up on time to pick up all your boxes? Did you finalize all your utilities and phone contracts?

Oh … and have I already told you? You're also taking yourself with you.

Look, I don't blame you for forgetting yourself in the chaos. It's only human to get completely caught up in this whirlwind of picking up your life, turning it upside down, reorganizing it, trying to make sense of it all.

Then, when you're at your peak state of confusion and overwhelm, it's time to board the plane and step into the unknown. Full of mixed emotions, you're saying goodbye to a life that once felt familiar and heading toward a new one that hasn't taken shape yet.

You're in the in-between. You haven't fully let go of the old yet, but you're not quite in the new either.

That's when the real challenge truly begins. You arrive at your next destination. Everything is new.

Maybe they speak a language you already understand. Great! That should make things easier. You feel confident at first, but then still you realize … the accent is different. You're struggling to understand what they're saying. You miss the nuances, the cultural context, the tone beneath the words. Connecting with locals suddenly feels harder than expected.

You also don't fully grasp the habits or social cues. Someone casually asks, "How are you?," and you think, how kind! So you answer, genuinely. But before you finish your sentence, the person has already walked off. Apparently, they didn't really want to know how you were doing. Weeks later, you learn that it's just a way of saying hello.

Story of my life.

You thought you prepared yourself for cultural differences. You read about the country, maybe watched videos, or asked others for advice. But soon, you realize: you brought your own culture with you. It changes everything. You're not a blank slate. You carry your own filter, your own lens, your own norms, your own way of seeing the world.

And in that, adjusting to another culture becomes a journey of self-discovery. You begin to notice how deeply you've been shaped by your home culture, more than you realized.

Of course, your host country comes with its own list of logistics, too. You need to register in your new city, sign a lease, open a bank account, get a new phone number, convert your driver's license, set up health insurance, find a local GP … the list keeps growing. Everything you worked so hard to cancel or rearrange before departure now needs to be built up again from scratch.

In that sense, it really does feel like starting over.

And that's what moving abroad often feels like: a fresh start. It's the promise of a new adventure, a chance to begin again. Something shiny on the horizon calling your name. The hope for a more fulfilling life, a career with more potential, a version of you that finally feels free.

Sometimes, it can also feel like an escape. Maybe you're done with the status quo. Maybe life hasn't unfolded the way you thought it would. You know that you deserve something better. You feel that pull inside you that says, "there's more out there for me." So you take a leap of faith to begin again somewhere else.

But here's the thing:

> Even though it feels like a fresh start … it's not.
>
> Your life doesn't reset. It continues. You're simply relocating yourself to a different place.
>
> But it's still you.
>
> You don't magically become a new person just by moving your body and belongings from one location to another. And it's not just your body that moves, you take your mental and emotional self with you, too.
>
> I don't say this to discourage you. Not at all!
>
> I've moved abroad several times now, and as I'm writing this, I'm preparing for my next move. And I can honestly say: these moves have been some of the best decisions of my life.
>
> I've also witnessed it with the clients I've supported since starting my coaching business in 2020. Not one of them has regretted their move abroad. It hasn't always been easy, but it's always been worth it.
>
> What I *do* wish I had known sooner though, and that I now want to share with as many people as possible, is this:
>
> Your relocation journey can be smoother, more fulfilling, and far less stressful if you also pay attention to what's happening inside of you.

Once you truly understand that moving abroad is as much an inner journey as it is an outer one, and you start living with that awareness, everything changes.

That's what I want you to take away from this chapter.

1.2 MY PERSONAL STORY: FROM IMMIGRATION CONSULTANT TO TRAVEL COACH

Before we dive into how you can make sure you don't forget yourself during the relocation process and how to stay connected to yourself when there are (literally) so many moving pieces, I want to share a bit more about my own journey so far.

Before I became a certified travel coach and life coach in 2020, I worked as an immigration consultant for almost a decade. I graduated with a Master's in International and European Law in the Netherlands, which included an exchange year in Spain. That was my very first relocation.

I was 20 when I moved to Barcelona, excited for the adventure but also a little nervous. I had never lived outside of my home country. Sure, I had moved out of my parents' house when I was 17 to study in another city two hours away, but this was different. Moving to a whole new country by myself was a much bigger stretch.

I didn't speak Spanish. I didn't speak Catalan. And soon after arriving, I discovered that my law classes at university were taught in Spanish. I shared a tiny apartment with three girls from Mallorca and Menorca who mostly spoke Catalan, but they kindly switched to Spanish when I was around to help me understand. At that point though, I could barely tell the difference between the two languages.

Since no one around me spoke English, I realized quickly: if I wanted this exchange year to work, I had only one option: learn Spanish fast!

So I threw myself into it. I sat in class with a dictionary, writing down whatever I could understand and translating on the spot, trying to make sense of it all and piece everything together. Let me tell you, when learning a language feels like a survival skill, you learn quickly.

I took my grammar book everywhere, even to the beach. I practiced speaking Spanish as much as I could. Thankfully, Spanish culture offers plenty of opportunities for socializing. And let's just say … after a glass of sangria, or a cocktail or two, you start to believe you're already fluent.

Fast forward six months, I was fluent, passed all my exams, and started to feel more like a local than an outsider.

So what helped me through those challenging first few months?

I wasn't afraid to fail. If I'd been too scared to speak or make mistakes, I never would have learned so fast. I knew not to take myself too seriously. I laughed at the weird things I said and embraced them as part of the learning curve.

Besides that, I was also fiercely determined. I really wanted to make it work. If it was possible, I wanted to make that year the best year of my life so far. I said yes to every social invitation. My motto back then was "I can sleep later." Now that I'm in my 30s I don't recommend that approach, but it worked for me at the time.

I went to cultural events and even became enthusiastic about soccer – something I never cared about before – just because everyone around me was so passionate about it. I loved the vibe, the energy, the sense of community. Now, 15 years later, my interest in soccer has faded, but I still appreciate how sports can bring people together.

A few years after Spain, and after finishing my law degree, I was invited for an internship at the United Nations Assistance to the Khmer Rouge Trials (UNAKRT) in Cambodia. I had a two weeks' notice. I packed my bags and, without much preparation, boarded a flight to Phnom Penh.

That was the first time that I truly understood what "culture shock" meant.

Again, I was alone. And because I had so little time to prepare, I honestly didn't know what to expect. The moment I stepped into a taxi from the airport to a hostel, where I would spend the first few nights while searching for a longer-term stay, I was hit with the reality of poverty, the chaos of traffic without any lights or rules, the flooded roads, the piles of trash. It was intense. I had traveled to many countries before, also in Asia, but I had never experienced anything like this.

Maybe I should have prepared more, I thought. But could I have, really? Some things you can only fully understand when they're right in front of you. You can read about it, learn from other people's stories and experiences, but nothing compares to living it yourself and being in the midst of it.

The months I spent in Cambodia were intense and shifted my perspective in many ways. Beyond the culture shock, I was deeply moved by the history I learned, and wondered why I had never been taught about it in all my years of school. My time working as a legal intern at the UN Tribunal taught me more than any textbook could have. I made friends with other expatriates and local colleagues, navigated language barriers, and found my way across cultures.

Looking back, that experience taught me something profound: the world is our biggest classroom. It's beyond the four walls of a school building that we learn who we are, what we're made of, and what life really has in store for us. Travel provides the opportunity to expand our horizons and turn us into global citizens.

After Cambodia, I returned to the Netherlands and began my corporate career as an immigration consultant. I helped expatriates through their visa processes and coordinated global moves for the international company I worked for. I started out in the Netherlands and eventually transferred to Australia. What was meant to be a temporary six-month assignment turned into a life-changing three-year adventure. More on that in a moment.

I worked with people from all over the world; different cultures, backgrounds, languages, and relocation journeys. Some moved solo, others moved with their families and pets.

What stood out to me, when I worked so closely with these expatriates, was how differently people responded to the process. Some people stayed very calm and confident, while others felt anxious and overwhelmed. And the strange thing was, I supported them in the exact same way.

That's when I realized: it wasn't about the logistics. It was about their internal process.

Why did one person barely ask any questions while another called me every day to check on their visa status, even though their timelines were practically identical?

Back then, I assumed it was cultural. Or maybe some people simply found it harder to let go of control. Whatever it was, it always intrigued me. Even though I wasn't a coach yet, I already knew that my job wasn't just about paperwork. It was about making sure people felt seen and supported as much as possible throughout the process. That's what I loved most: the human side of it, beyond the logistics.

My first stop in Australia was Sydney. I'll never forget seeing the Opera House and Harbour Bridge for the first time. It felt like I had stepped into a postcard or a National Geographic documentary. Australia had always felt like a dream. A faraway dream, an unreachable bucket list destination.

And yet there I was. I made it happen. And once again, I was determined to make it the best time of my life. While I was deeply immersed in my fulltime job, where I often worked 50-to-60-hour weeks, I still pushed myself to explore. Even when I was exhausted after a long day at the office, I'd go to social meetups, walk the city, or dance at a local salsa night. on the weekends, I made it my mission to visit beaches and go on hikes through national parks. I wanted to experience it all.

But after a few months, I hit a wall. I couldn't keep up the pace of this lifestyle. I realized that burning myself out just to "make the most of it" wasn't sustainable. So, step by step I began to set boundaries, listen to the signs of my body, and take breaks whenever I needed. I took my self-care seriously for the first time.

After six months, my visa was about to expire and it was time to move back to the Netherlands. But I didn't want to leave. I wasn't ready to close that chapter yet.

Thankfully, my employer sponsored me and I was able to extend my visa. My visa deadline was moved into the future again and I had space to breathe.

Living with a visa deadline is a strange feeling. Life is always uncertain, but when your stay in a country is defined by a literal expiration date, it changes how you approach each day. It changes your mindset. You become more aware of how precious your time is, and you become more grateful for each day. At least, that's how it felt for me.

I didn't want that looming deadline to define my time abroad. I wanted to use it as a reminder to stay present and make the most of each day, because you never know how things will unfold.

After a year in Sydney, I got promoted and was transferred to the Melbourne office. Even though I truly loved living in Sydney, I hadn't been to Melbourne yet, and the idea of a new chapter, a new city, and a new rhythm excited me.

By then, I had met my now husband (who's Portuguese, not Australian), and we decided to move to Melbourne together. We signed a lease for an apartment on the 59th floor of a 91-floor skyscraper in the heart of Melbourne CBD. Every morning, when I opened the curtains, I was blessed with a breathtaking view of the city and the mountains in the background. Coming from the flat landscapes of the Netherlands, seeing mountains every day and living in between the clouds felt unreal. I never took it for granted.

This was the chapter in my relocation journey in which everything started to make sense to me. I overcame the initial challenges, had found my flow in my career, made deep friendships that felt like family, and shared a home with the love of my life. I felt like I had cracked the code of a successful expat life.

What could possibly go wrong?

Then … 2020 happened. The pandemic hit, and Australia closed its borders. No more expatriates were arriving. Most of the international moves I was managing came to a halt. My job was made redundant, and I lost my visa.

Within two weeks, I had to pack my bags and leave the country. Leaving a place that felt like home was one of the hardest things I've ever done. Especially because it wasn't by choice.

The world was in lockdown. We had no idea where to go. Eventually, my partner and I decided to move back to Europe and temporarily settle in the Netherlands. One day after we left, Melbourne's international airport closed.

Those last two weeks in Melbourne were a blur. I was grieving the loss of my job and flourishing career, and even more, the loss of my home. We had to make quick arrangements while most services were closed. Due to the strict Covid-19 measurements, we weren't allowed to meet anyone. We couldn't even say goodbye to the friends who had become our family. We packed everything into a container and shipped it overseas, because there was no time or possibility to sell, donate, or give things away.

The first six months in the Netherlands were even blurrier. I wasn't myself. I cried every day. I felt lost, disconnected, depressed, confused, purposeless, and on top of it all, stuck in another lockdown.

Moving back to my home country was the hardest move of my life.

And I experienced something I didn't know existed: reverse culture shock. It's a culture shock in a country that you're already familiar with.

How could it be this hard to return and adjust to the country I grew up in? I spoke the language. I knew the culture. I knew my way around. And yet, I felt like a stranger.

What made it even harder was that no one around me truly understood what I was going through. And how could they? If you've never lived abroad, you can't fully grasp the layers of grief, identity shifts, and loss that come with coming "home."

Even though I had been dipping my toes in self-growth for a while, I knew that I had to go deeper to get myself out of the funk I was in. I didn't want to feel like a victim. I was still healthy. My partner was with me. I had a deep inner trust that I'd find a way through.

And that strong determination that I always carried with me abroad? I had to reconnect with that again. That was at least a silver lining.

So I went inward. I started focusing on my personal growth, healing, and recharging. I decided that it was time to take control of my life. I had always dreamt of starting my own business, and during this period I discovered exactly what I was meant to do.

I got certified as a travel coach and life coach. And I started helping others on their personal journeys. Not just with the outer journey, but with the inner journey too. Because those two paths? They cannot exist without each other.

I'm grateful that since taking that leap, I've supported clients on incredible journeys. One client decided to leave the United States to study in Europe, and found love. Another client moved to the United States, lost his job, pivoted his career, and found work that truly lights him up. Another client decided to extend her stay abroad and found a better balance between her work and private life.

At one point in my coaching journey, I had an Australian client trying everything he could to stay in the United States, an American client doing everything to leave the United States and move to Europe, and myself being in Europe, missing Australia and looking for ways to move back.

It was the perfect paradox and a great reminder that relocation is very personal. There is no one-size-fits-all (or should I say "one-country-fits-all") when it comes to the places where we feel called to. It's subjective. It's layered. And it's always evolving.

1.3 THE IN-BETWEEN PHASE

No journey and no relocation story is ever the same. However, over the years of working with people from all around the world, I've noticed that no matter where they're from or where they're headed, there's one phase we all go through.

I call it the in-between phase.

This is the space between the decision to relocate and the actual move. You're still living your current life, but you've mentally and emotionally started letting go. At the same time, you're envisioning a future that hasn't started yet. It's a limbo, between the familiar and the unknown.

This is the phase where you're preparing to invite something new into your life and while leaving behind something old and familiar.

Even though the idea of a fresh start feels promising, there's always a part of us that will come with us wherever we go. We often completely forget about that part. It just tags along quietly. Like the air we breathe. We don't really notice it until something happens that brings it to the surface.

Perhaps you've been aware of your mind. Maybe you've already noticed your mind racing with thoughts. When you're in this in-between phase, or even when you're just dreaming of moving abroad, your thoughts can be all over the place.

One moment, you think about all the possibilities and you're full of excitement. The next moment, the only thing you can focus on is all the obstacles that you may need to face to make it happen. One moment you have total clarity and know exactly which next steps to take. The next moment you feel overwhelmed by all the things that could go wrong and you're lost in a sea of logistics and uncertainty. One moment you're confident about your decision to move. The next moment, you doubt everything and you wonder if it might be easier to just stay where you are.

But this in-between phase goes deeper than just what's happening in the mind.

While your thoughts are playing out all kinds of "what if" scenarios of what possibly could go right or wrong, your body is processing emotions that come with them. Stress, anxiety, fear, uncertainty. These don't just live in your head. They live in your nervous system.

If we're not aware of that, we tend to suppress those uncomfortable feelings. We think we don't have time to deal with them, or we just don't want to feel them. So we push them aside, and they get stuck in our bodies.

When you're about to make a leap as big as moving countries, it's completely normal to feel a mix of fear and excitement. Fear and excitement are actually very closely linked. Often they only require a mindset shift to move from one to another. The bodily sensations we associate with both fear and excitement – like adrenaline, a racing heart, sweaty palms, or nervous energy – are basically the same.

The difference is how we frame those sensations. Is it danger? Or is it adventure? Is it uncertainty? Or is it possibility?

These emotions show up when we don't know what to expect, for leaving our comfort zone behind, and for feeling uncertain about our life choices. That's where mindset comes in. Depending on our mindset, we can see this as a threat or as an exhilarating adventure.

Beyond fear or excitement, other emotions may show up during this phase as well. When you think about leaving behind the familiar, sadness is common too. Even when you're making a conscious choice to move, a part of you is still grieving what you're leaving behind. We don't talk about this enough. We often tend to focus on the exciting parts of moving, focusing on the future and the new life that's ahead of us. However, it's just as important to give yourself space to process emotions like grief, confusion, or guilt.

For some, sadness comes from saying goodbye to a familiar place without exactly knowing what you'll find on the other side. For others, it's the pain of leaving loved ones, perhaps not knowing when you'll see each other again or how distance will change your relationships.

Confusion can creep in as well, especially around your identity. While you're in the process of becoming your next level self, the one who had the courage to take the leap and say yes to this move, at the same time, you're also shedding old parts of yourself that no longer fit. The in-between phase can be a turning point in which you reflect on who you were, who you are now, and who you are becoming. This is a beautiful transformation, but it can also leave you confused when your inner foundation is shifting.

And then there's guilt.

Guilt shows up when we ask ourselves, "Shouldn't I just be grateful for the life I already have?," or "Is it selfish to want something more?" Maybe you feel guilty about leaving loved ones behind. If you've asked yourself these questions, know that you're not alone. In fact, it shows how much you *do* care, and it doesn't mean that you do not deserve to take this step.

Taking time to sit with these emotions during the in-between phase is key to how you'll integrate into your new life later on. If you understand what's showing up and give yourself space to process it, you set yourself up for a smoother, more grounded transition. This phase is the initiation for the leap you're about to take.

Let's take guilt as an example.

If you already feel guilty before you move, about being far away from your family for instance, that guilt can become even heavier once you've left. Every time you talk to them, it resurfaces. And when it does, it may prevent you from fully enjoying your life abroad. Deep down, you carry this belief that you don't deserve it. That you're selfish for choosing your dreams over proximity to others.

But what if you could reframe that guilt?

What if you sat with it and explore why it's there? What if you gave yourself compassion and grace instead of judgment?

Moving abroad doesn't mean you love your family or friends any less. It means you're honoring a calling within yourself. It means that you give yourself permission to live the life you've always dreamt of. It means you trust that by choosing a life that fulfills you, you're also becoming someone who has more love, presence, and purpose to give back.

People who love you want you to be happy. Even if that means being further apart. Both truths can co-exist at the same time.

Ultimately, it comes down to this:

You can choose a thought that keeps you stuck in victim mode, guilt, fear, or self-doubt. Or you can choose a thought that supports and empowers you in your journey.

I know that this sounds simple in theory. But in practice, it's not always that easy, because of our self-talk, conditioning, limiting beliefs, and external input. That's why this isn't a one-time decision. It's a practice. It's an ongoing process of checking in with yourself and consciously choosing how you want to feel.

And it starts here, in the in-between.

1.4 SETTING THE STAGE FOR A SMOOTH MOVE

So how do you actually do that?

How do you focus on yourself when there are (literally) so many moving pieces?

Is it even humanly possible to *not* feel overwhelmed when your entire life is packed into boxes?

When you're saying goodbye to people you love, not knowing when you will see them again?

When you're at the airport, passport and boarding pass in hand, waiting for your flight, wondering if you made the right decision, and having no clue what will happen once you land?

Let me be honest with you.

Even with all my experience – even after coaching so many clients, moving multiple times myself, receiving coaching, therapy, and healings, working with mentors, and building a solid toolkit of practices – I still get overwhelmed and stressed when I move. I still feel it. Every time.

And that's okay.

This journey is not about eliminating those feelings altogether. They're part of being human.

What matters most is that you learn to recognize them. That you become aware when stress, fear, or anxiety shows up, so you can acknowledge it and work with it.

This is what is often the hardest part, honestly. We were never really taught how. It's not something we've learned at school. It's usually the opposite, as kids we were often told not to cry. We learned to suppress emotions because they made others uncomfortable, or made us feel weak. Now, as adults, we still carry that story. If we cry in public, we apologize or try to hide it as much as possible. We've learned to "be strong," which often means pretending we're fine.

But here's the truth:

> You become stronger when you allow yourself to feel. When you're able to express your emotions, instead of pushing them down. When you become aware of what's going on inside of you, sit with it and take care of yourself.

I know this can feel very uncomfortable, especially if you're not used to it. And even if you are someone who feels in touch with their emotions, I invite you to ask yourself:

> "How often during this relocation journey have I let myself become consumed by my emotions, feeling completely overwhelmed or stressed, without really knowing how to pull myself out of it?"

Let's pause, and let's come back to *you*.

How would you like your relocation journey to unfold?

With chaos, stress, and emotional shutdown?

Or with calm, clarity, and a sense of inner steadiness?

Now, I can't promise you that there won't be any challenges. Life happens, after all.

But what I *can* promise is this:

> When you focus on your inner journey as much as you focus on your outer journey, there's a much higher chance of a smoother, more grounded move.

A "smooth move" goes beyond perfect logistics. You can have everything organized and still feel stressed or overwhelmed. What you're feeling is your nervous system reacting to change. Your brain is wired to protect you, and whenever you stretch yourself and step out of your comfort zone (like moving to a new country), your internal protection mechanisms start to raise alarm bells as if you're in danger. Your body goes into alert mode. This protection system is useful when we're actually in danger. But in everyday life, especially during a move, those same internal systems are often outdated or overactive.

A move abroad is a big leap into the unfamiliar. Your internal "safety team" will work overtime and is on high alert to manage all the new triggers that come their way, scanning for anything that feels risky.

It's important to understand that this is normal. It's how your brain operates. What matters is how you respond. You don't have to be ruled by those reactions. There are ways to avoid this from holding you back or feeling like you have no control over your state of being. You can create a space between the trigger and the response. That's where self-awareness comes in.

A smooth move starts when you can calm your system, re-center when challenges occur, and regain control of your inner world, even when the external world feels a bit chaotic sometimes.

Let's start with understanding how our mind works and becoming aware of our thinking patterns. Your thoughts create your experience.

When it comes to moving abroad, one thought shows up and this one thought can branch out into many directions. Depending on what you choose to focus on when you think about your upcoming move – whether it's a fear-based or empowering thought – will directly affect how you feel. How you feel will determine what action you take, and that action creates your outcome.

For example:

> In the case of moving abroad, if you're led by fear, your choices might come from a place of stress or avoidance. But if you're led by excitement or grounded intention, your actions feel more aligned and intentional, even if the exact same task is in front of you.

The move itself may still happen either way, regardless of whether you choose thoughts that empower you or overwhelm you. But your *experience* of that move will be completely different.

In one version, you see every obstacle as a crisis, leaving you exhausted and overwhelmed. In the other version, you see it as a challenge you can handle, navigating the process with clarity and confidence.

That's the difference between a move that leaves you drained and one that leaves you feeling empowered and ready for your next life's chapter.

I know, you might be thinking: "But I can't control my thoughts. They just happen!"

You're right, most thoughts happen automatically. We have thousands of thoughts each day. It would be exhausting and impossible to track every single one of them.

But you don't need to. All you need to do is become aware of your thinking patterns and observe them.

For example, let's say you notice that every time you have to gather documents for your visa process, you get tense and overwhelmed. Instead of pushing through, pause, and observe where this is coming from.

Ask yourself: "What's the thought behind this feeling?"

Maybe it's:

> "What if I can't get this document in time, and I miss my visa deadline?"

That's a fear-based pattern. It makes you feel anxious, frozen, stuck, or out of control.

Now, let's try shifting this thought pattern into a different one:

> "I don't know how to get this document yet, but I'll figure it out. I trust that the support I need will show up."

It's the same situation, just a different choice of thoughts and a different energy. However, your experience will be completely different.

You still have to do the same thing, but now you're doing it with a calmer nervous system and a clearer mind.

When you become aware of your thought patterns, you give yourself the power to choose differently. It becomes the first step toward feeling empowered for a smooth move abroad.

Once you realize that you're not your thoughts, that you can step back and observe them, you gain access to a deeper truth:

> You have the power to respond, not just react.

We can choose different thoughts to create our current reality. And this becomes the difference between surviving your move … and thriving through it.

There's still a deeper layer that we can look at though; let's dive in.

1.5 FINDING HOME WITHIN

In my experience, this is where it all begins. While you're physically changing locations and searching for a new place to call home, whether that's an apartment, a townhouse, or a villa, there's also another home to consider:

The home within yourself.

We often define home as a physical space, a building with bricks, windows, and a front door. But I believe home is more of a feeling. It's the place where we feel safe. Where we can unwind after a long day. A space where we can simply be ourselves.

And I want to take that idea even a step further:

> What if you could access that feeling of home, no matter where you are in the world? Wouldn't that be powerful?

> Wouldn't it give you a sense of ease and calm to know that even though you're stepping into the unknown and don't know what to expect, you know for sure that whatever happens, you can always come home to yourself?

I think it's a beautiful thought to consider. That thought alone gave me so much peace of mind when I discovered this for myself. Because the truth is that I had always been looking for a place to call home. I've loved travel since I was a kid, and somehow the grass always seemed greener somewhere else. And honestly? That turned out to be partly true for me. Some of the happiest years and moments of my life have happened while traveling and living abroad.

But there's a nuance here.

At some point, I realized that I also needed to feel at home in the Netherlands, my birth country. I didn't want to only feel good when I was traveling or living elsewhere. I didn't want the feeling of home to depend on a location. I wanted to be able to tap into it wherever I was, especially in those moments when I felt disconnected or uncertain and missed a sense of belonging.

And I want that for you too.

I want you to belong to *yourself*.

To feel at home *within* yourself.

So that no matter what happens during your relocation journey, when things get messy or you feel stressed, disempowered, lost, or lonely, you know that you've got your own back. You can count on yourself. You can support yourself.

You feel grounded in your daily choices and in the way you move through the world. It's a steady inner foundation. An inner knowing that you can handle whatever comes your way.

That's what keeps you going, one step at a time, when you're navigating language barriers, cultural misunderstandings, or trying to build meaningful friendships from scratch.

But this kind of grounding doesn't come from a checklist. It comes from something deeper: self-trust, self-awareness, and self-love. Yep. Big words, I know! You thought this chapter was about preparing to move, and now we're diving into these inner layers. I get that it might feel a little uncomfortable.

But if you're ready to make this move the best decision of your life, and if you want your life abroad to be one of the most beautiful chapters yet, keep reading. You won't be disappointed.

What I've seen in my own journey, and in many of my clients, is that we're often searching for a sense of belonging. We don't feel like we belong in our current location anymore, so we hope to find that feeling elsewhere.

Belonging has many layers. It can come from being part of a group, a community, a circle of friends, a family, or a romantic relationship. It can also come from feeling connected with something bigger than yourself, whether that's the universe, nature, spirit, or whatever resonates with your belief system.

But at the heart of it all, belonging starts with you. It starts with the connection to yourself.

If we only seek belonging in others, we're constantly reaching outside of ourselves, hoping that someone else will meet our needs. That they'll make us feel seen or understood, and without them, we risk feeling lost or alone.

But guess what:

> You can be surrounded by people and still feel lonely. You can feel lonely even when you're not alone.
>
> It works the other way around too. You can be completely alone and not feeling lonely at all.
>
> When you're happy with your own company, when you know how to nurture your inner home, you create a sanctuary that no one can take from you.
>
> That's the place where you're so content that no matter what's happening in the outside world, you can always return here.

And if you can connect with this inner home before your move, the challenges that come after, while adjusting, settling in, or feeling out of place, will become much easier to navigate.

1.6 THE INNER PREPARATION FRAMEWORK

Let me share a simple framework that can help you stay grounded through the messiness of your move and the chaos of the post-arrival phase. It's designed to help you come back to yourself, again and again, no matter where you are.

1.6.1 Step 1: Know Your Why

First of all, I want you to tune into your *why*. Why do you want to move abroad?

Really sit with that question, tune into that for a moment, and focus on what the outcome is that you want for yourself. What is it that you're truly looking for? What outcome are you hoping this move will bring?

This step is so important because often we move for reasons that seem logical or exciting on the surface. Maybe the grass looks greener elsewhere. Maybe you're craving adventure. Maybe your current life feels boring, limiting, or no longer aligned. Maybe you're frustrated with the political or economic climate of your home country and just want things to feel easier.

Maybe all of that is true for you. But what's underneath it? What are you *really* missing in your life? What's the deeper need behind your desire to move?

Instead of focusing on what you want to "escape," try focusing on what you want to create.

When you think about your why for your move, tap into this place of belonging to yourself.

Ask yourself:

- "If I wasn't worried about what other people think…"
- "If I didn't feel pressure to do what I'm 'supposed' to do…"
- "What would the real reason be for this move?"

What outcome are you hoping to experience, just for you?

Take time to reflect. Journal on these questions. Meditate. Go for a walk. Dance it out. Do whatever works for you to connect with that deeper part of yourself.

Why is this WHY so essential?

Because when things get hard or stressful (and at some point, they will), this is the anchor you'll return to. This is the reason why you're taking this step. This is the foundation of your entire journey and everything you will be doing next. It reminds you why you're doing this in the first place. This is your intrinsic motivation, and it matters more than any checklist or packing strategy ever will.

1.6.2 Step 2: Tune Into Your Body

Secondly, once you've found more clarity around your *WHY*, I want you to check in with your body. When you think about your move, what sensations show up?

Most of us let our minds run the show. That's not necessarily a bad thing, because your mind helps you stay organized. As we previously discussed, you now know that you have the power to choose more empowering thoughts. However, you're not just your mind. You also have a body, which wants to be included in the conversation too.

Your body is the place where your emotions land after your thoughts arise. If you're not tuned in, you'll miss the signals your body is sending you. While your mind and body run mostly on autopilot, the signals help you regulate your nervous system and stay grounded.

So pause for a moment. Take a deep breath in. And if it feels good, place one hand on your heart and one on your belly.

Ask yourself:

- How does my body feel right now?
- Is there tension somewhere?
- Am I breathing deeply or shallowly?
- Is my heartbeat calm or racing?
- Are my shoulders relaxed or tight?
- Am I clenching my jaw?
- Do I feel tired or energized?
- Do I feel scattered or grounded?

Checking in with your body, even just a few times a day, can be a game-changer. Especially during the chaos of a big move.

When I said that we often forget ourselves during relocation, the body is one of the most overlooked areas.

Believe me, I've been there too. I spent most of my life living in my head. Sure, I ate healthy, exercised, and got enough sleep (well … if we skip my twenties). But I didn't truly understand how much stress and tension my body was holding until I hit a wall and had to stop. I felt exhausted, burned out, and disconnected.

Don't let it come that far. Don't wait until your body shuts down to listen to it. Burnout is more common among expats than we realize, and we don't talk about it enough.

I'm not saying this to scare you, but to remind you: even if you think that you have everything under control, are high performing, and used to pushing through, your body needs care. Your inner home includes the vessel that carries you.

Now that you have this framework, you don't just have to read it, you can actually live it.

If you're willing to apply this framework to your life, you will notice the difference and you'll avoid falling into the trap of "I don't need to rest" and "I can just push through."

Next time when you feel tension or stress in your body, try something new:

Start a conversation with it.

I know how strange that may sound, but just give it a try. Yep, talk to your body.

Ask your body: what do you need right now?

And notice what answers show up.

It might be a nap, it might be a walk, or some fresh air, or a warm hug, or just a moment to breathe with a cup of tea.

Whatever it is, listen to it. This is the simplest, most essential form of self-care.

The more you listen to what your body needs, the more regulated your nervous system will be. The more regulated you are, the calmer your mind becomes.

Because just as your mind influences your body, your body influences your mind too. When you calm one, the other follows. Make them part of the same team.

1.6.3 Step 3: Build Your Personal Toolkit

The final step of this framework is all about support. Now that you have connected with your why and have built a stronger relationship with your mind and body, it's time to create your personal "emergency" toolkit: your go-to resources when you feel stressed, overwhelmed, or unsure during your move.

You can think of it as your inner carry-on. Just like on a flight, if your carry-on is too heavy, you have to check it or pay extra. It slows you down. It literally weighs you down. It makes the whole journey harder.

Your inner carry-on works the same way. If it's filled with old fears, outdated beliefs, or pressure to be perfect, it's going to feel heavy. It will drain you. But if it's light, intentional, and filled only with what supports you, it's easy to carry around and you'll move around more freely.

So, what do you pack in your *inner* carry-on?

See it like packing your luggage, and include anything that helps you reset, recharge, and reconnect. Create a toolkit that is filled with practices, habits, and rituals that support you and empower you. Whatever you need for a quick reset, to recharge, or unwind.

Maybe it's:

- A playlist of songs that lift your mood.
- A calming meditation or breathwork practice that grounds you in the present moment.

- A scent that brings back good memories.
- A piece of jewelry that reminds you of someone you love.
- A photo, plush item, journal, or candle.

It doesn't have to be anything big or complicated. Start by looking around. What do you already use every day that helps you feel good? What helps you come home to yourself?

Even something as simple as putting on your favorite song and dancing around the room can shift your mood instantly.

This toolkit is completely yours. Make it intentional. Make it accessible. Have fun with it. Let it evolve with you. Because the more resourced and resilient you feel on the inside, the smoother and more empowering your journey abroad will be.

1.7 THE POWER OF UNPACKING

The Inner Preparation Framework helps you to build a strong inner foundation. It gives you the tools to comfort or ground yourself and to come back to your center whenever life feels like a rollercoaster.

But while the final step of the framework focuses on packing your inner carry-on with grounding rituals and comfort items that support you throughout your journey, I also want to take a moment to talk about something just as important:

Unpacking.

My coaching business is called *Unpackbyveer*, and the name is rooted in this exact metaphor. I came across this concept when I started asking myself:

How can we travel lighter through life?
What are we still carrying that we don't actually need?

We often hold on to unhealthy patterns, beliefs, or unresolved emotional baggage for years, sometimes decades. Old pain, outdated habits, limiting beliefs; when we don't process or release those things, they become heavy.

We wonder why we feel tired, easily triggered, or emotionally drained, disconnected from ourselves or others. It's because we're still carrying things we were never meant to hold forever, and our inner luggage becomes heavy.

That's why unpacking is so important.

For me, unpacking means getting to know yourself on a deeper level. It's the process of noticing what you're carrying that no longer serves you and letting it go. In other words, it's about releasing the emotional and energetic weight that's been slowing you down and that doesn't support you.

Yes, this can be uncomfortable and confrontational. It requires courage and curiosity. You have to visit the corners of your mind and heart that you may have avoided for a long time. However, if you allow yourself to create space, something beautiful happens. It not only opens up room for new experiences to come but also supports you to move lightly through life with more joy, confidence, and clarity. Most importantly, there is space for the real *you* to come forward.

As a coach, this is the deeper inner journey I support my clients with, because moving abroad isn't just about physically relocating. You will move in all of your dimensions: physically, emotionally, mentally, energetically.

If you push parts of yourself away or leave certain parts of yourself behind, you'll eventually feel it. You'll notice that you're living in two worlds at the same time: the version of yourself you left behind, and the one you're trying to become. It can become a struggle between an old identity that no longer fits and a new one that you haven't really fully integrated yet.

Without integration, you can feel internal conflict. This is when feelings like loneliness, homesickness, or lack of purpose may show up, even when everything around you looks great on the outside.

That's why this work matters so much. Unpacking helps you close the gap. It helps you to let go of what no longer fits and to step fully into the identity you're growing into.

The goal isn't just to adapt to a new country. The goal is to feel at home in yourself, so you don't end up feeling lost. Not just literally lost in your new city, but lost within yourself.

When you unpack what's weighing you down and embrace your full, authentic identity, you become whole again. You move forward as the most grounded, true version of who you really are.

1.8 PREPARING FROM THE INSIDE OUT

Trust your inner GPS and prepare yourself for the changes ahead by staying present in the moment and getting crystal clear on the *real* reason behind your move. Not just the surface-level logistics or the checklist tasks, but the deeper personal layer.

What is your true intention? What's your purpose? Why are you doing this, not just in theory but in your heart?

If you're not clear, you might end up chasing a version of success or freedom that doesn't even belong to you. You chase a goal that never really fulfills you or a dream that isn't truly yours.

It's like the person who travels to dozens of countries, always chasing the next thrill, hoping the next place will finally bring peace, meaning, or satisfaction, only to find himself emptier than ever before. What is this person really looking for? Maybe they're just trying to outrun something, distracting themselves from getting too close to the real them?

The outer journey always includes an inner journey. In fact, the inner journey might even be more important, because it influences everything else you do.

So let me ask you:

When was the last time you checked in with yourself?
When was the last time you looked inside your inner carry-on?
Are you taking unnecessary things with you on your trip (your life) that are weighing you down?

Are you carrying things that aren't even yours (old beliefs, pressure, expectations from others)?

Do you even know what to bring with you, what matters to you, outside of what you've been told you "should" do?

This is your invitation to reflect.

What do you really need to feel ready for the next step in your journey?

What goes deeper than your visa, your budget, and your paperwork?

Who is the version of you that is going to take this step?

As I said, this is a journey in itself. You don't need to have all the answers to these questions right now. They'll come in layers. They'll evolve with time. Have patience with yourself. There's no rush. Your life's journey doesn't begin with moving abroad, and it doesn't end when you arrive. Relocating is just one chapter. Life continues far beyond it.

Through every chapter, one thing remains true:

You take yourself with you.

So take your most empowered version with you. Know that wherever you go, you can always come home to yourself. Remember: you don't have to do this alone.

Find people who can support you, both on your inner and your outer journey. When you're clear on who you are, you'll also know who you want to connect with, and what kind of relationships you want to build.

The more grounded and present you are, the more open you'll be to meeting new people, and the safer it will feel to be seen, to be vulnerable, and to build a truly beautiful community abroad.

At the same time, it becomes easier to stay connected with loved ones back home. When you're centered, you can create a balance between life in your host country and life in your home country, without feeling like you're being pulled between two worlds or identities.

You already have everything inside of you to make this move a success.

Trust yourself. You got this!

1.9 FINDINGS

In this chapter, we've taken a deep dive into the one thing people often forget when moving abroad: themselves.

Wherever you go, you take yourself.

However, often we're not fully aware of what's happening in our inner world, so we end up bringing a version of ourselves that isn't fully supported or empowered.

Our inner carry-on is filled with fear-based thoughts, limiting beliefs, or old conditioning that no longer fits who we are today.

That's why a smooth relocation isn't just about logistics. It's about who you are while you're moving.

When you become aware of your thoughts, you can begin to shift your thinking patterns. You start choosing more helpful thoughts that support you throughout the relocation process, instead of sabotaging you. This isn't something you do just once, but it becomes a practice that strengthens you for the rest of your life.

Through *The Inner Preparation Framework* and the *Power of Unpacking*, your move can become much more than a physical relocation. It becomes the most beautiful success story. A transformation from the inside out.

The Inner Preparation Framework is a three-step process to support your emotional and energetic readiness:

1. Focus on Your Why: Reconnect with the deeper motivation behind your move.
2. Tune Into Your Body: Listen to what your body is telling you beyond the mental noise.
3. Build Your Personal Toolkit: Create supportive practices and rituals to ground and empower you.

The Power of Unpacking is about letting go of stories, habits, and patterns that no longer serve you. It's about making space in your inner carry-on. Releasing what's heavy so you can move through life lighter, with more clarity, confidence, and freedom.

1.10 CONCLUSION

I know that if this kind of inner work is new to you, it can feel like a lot to take in. And even if you *do* have experience with working on your mindset or personal growth practices, you already know that there's always more to uncover.

To be honest, I still find myself spiraling down a rabbit hole from time to time, especially when I'm preparing for my next move abroad.

But here's what matters:

> It's not about reaching a point when you never feel stressed or overwhelmed again. We're human after all! All of our emotions are welcome and are a necessary part of the journey. They're not here to be avoided, they're here to teach us. They show us what we want and don't want, what we value, what we're ready to let go of, what empowers us, and what still needs healing. They're like signposts, gently guiding us along the path of our lives. Sometimes we take a detour. Sometimes we end up on the scenic route. And sometimes, we take the high road. No matter the route, every step is part of your becoming.

My hope is that this chapter offered you a supportive nudge. A reminder that you already have everything that you need inside of you to make your move abroad a success. That no matter how messy or chaotic things get, you can always come home to yourself. That you can offer yourself grace and compassion, especially during the moments of your relocation journey that don't go as planned (because those moments will happen).

When it feels like everything is falling apart, you can trust that you can get back up and that you're not alone.

I hope that you create the most beautiful inner carry-on for all your future travels. A personal toolkit filled with joy, empowerment, clarity, and confidence.

You're not your circumstances. You're more than anyone can see on the outside. Your past experiences don't define your future. You are someone who chose the leap. You left the familiar behind to step into the unknown.

You chose courage over comfort. You chose purpose over pain. You chose resilience over complacency. You chose excitement over fear.

And I'm here cheering you on.

AUTHOR'S NOTE

Writing this chapter brought me back to so many moments in my own journey; the airports, the boxes, the meltdowns and the breakthroughs, and all the quiet in-between phases where I had to learn how to come home to myself. Again and again.

To my coaching clients: thank you for trusting me with your personal journeys and allowing me to witness your courage and growth. Your stories inspired this chapter and motivated me to share my message with a broader audience. I'm grateful to walk beside you as you build your next chapter, wherever life takes you.

And to you, dear reader: thank you for being here. For choosing this chapter. For being open to the deeper layers beneath the move. And for trusting that the path you're on matters.

You don't need to have it all figured out. You just need to stay connected to the part of you that already knows the way.

If something in these pages resonated with you, whether it sparked a shift, soothed a doubt, or simply reminded you that you're stronger than you think, I'd love to hear from you.

You can visit my website at www.unpackbyveer.com or send a message to info@unpackbyveer.com. I'd love to hear your stories and support you on your journey, wherever you are, and wherever you're going.

With love and travel vibes,
Veerle
Veerle Beelen Da Costa David

2 Thoughts on Reinventing Yourself as an Expat

How Internal Resources Support Women Relocating to the USA

Izabela Jastrzebowska

2.1 INTRODUCTION

Life's circumstances have taken me away from California when I decided to write this chapter about my expat's journey in the USA. This journey was possible because of the curiosity, resilience, flexibility, and out-of-the-box attitude toward life's events, which were among my most precious internal resources. They constitute a healthy compass with a needle aligned with my values.

In this chapter, I share my remarks about challenges, myths, and joys of expatriating to the USA. I encourage you to seek your own life path. I share a message of support. Realize your own fulfilling story, not despite the difficulties, internal limitations, external influences, and expectations, but thanks to them. Use your omnipresent internal guidance which tells you which way you should go, one step at a time.

I am a Jungian-oriented psychologist. In practical terms, being Jungian means that, like many of us, I easily connect to my right brain hemisphere, which means internal world, senses, feeling, intuition, dreams, and perception of various aspects of life as interconnected. To be Jungian at heart means to embrace Carl Jung's philosophy of self-discovery and interconnectedness. It reflects a mindset that values introspection, intuition, and the exploration of the unconscious mind. Being Jungian involves recognizing the significance of dreams, symbols, and archetypes as guides to understanding oneself and the world. It also emphasizes the idea that all aspects of life—emotional, spiritual, and rational—are deeply connected.

In its essence, being Jungian is about striving for ***individuation***, Jung's term for ***becoming*** the truest, most integrated version of yourself. It means honoring the interplay of light and shadow within, learning from your inner world, and approaching life's mysteries with curiosity and openness.

 DOI: 10.1201/9781003604402-2

In this chapter, you will learn my personal story of expatriation and how it intersects with the broader, universal processes of becoming a whole person. This journey is closely tied to my transformative experience abroad, which ultimately led me to stronger self-awareness and international work in the fields of psychology, coaching, and arts.

2.2 DREAM OF CALIFORNICATION

> Don't confuse tourism with immigration.
>
> —a friendly advice

On a windy Tuesday, January 17, 2017, I landed at San Jose Airport in California, unaware of what I had left behind and what was about to unfold before me. Silhouettes of palm trees struggling against strong winds and pouring rain, me holding the small hands of my two baby girls, and my husband packing six suitcases into the massive trunk of a black car—this is what I remember from my first day in the Silicon Valley.

This vivid memory naturally reminds me of a family story about my great-grandparents, who came to America in the 1920s aboard a transatlantic ship. The interior of the ship was designed by an avant-garde Polish artist, Prof. Wojciech Jastrzębowski, a member of my husband's family. After two years, my great-grandparents decided to return to Europe—a decision as courageous as their journey to America.

When I recall those days, I think of a friend whom I regard as a "magical helper" put on my family's path. The best piece of advice I ever received came from George, who had relocated to the Silicon Valley shortly before us. He "warned" us not to confuse California's touristic beauty and the illusion of the California Dream with the reality of daily life there.

The first signs of "George's prophecy" began to materialize about three to six months in, when we started to lose the corporate safety net that had been supporting our family. This marked the end of our "starter package," which meant no more financial or logistical assistance with securing an apartment, furniture, salary, or car. From that moment on, we were entirely on our own, and the honeymoon phase of our relocation came to an end.

It took another three years of settling in for me to fully understand what George truly meant by those words. I had begun to understand that to expatriate is to more or less consciously release the familiar version of yourself, embrace the uncertainty of the journey ahead, and learn to find joy in the risks of inner transformation. In her book *Necessary Losses*, Judith Viorst (1986) explores the many aspects of what we must let go of in order to create space for the new. For me, "New" (with a capital "N") represents all that life, fate, God, the Source, or the creative force—depending on the religious or philosophical tradition closest to the reader—brings into our lives. Personally, I feel drawn to a Christian tradition intertwined with Eastern philosophies, transformed by modern psychology—they provide fertile ground for the consulting tool kit I use to coach myself and others daily.

2.3 DEEPER PERSPECTIVE ON CHALLENGES OF STARTING FROM SCRATCH

> One who looks outside, dreams. One who looks inside, awakens.
>
> —Carl Jung

Coming to America required a great deal of curiosity, flexibility, open-mindedness, spontaneity, naivety, and a willingness to take risks on my part. I chose to leave behind the stable income of my previous career in the legal field, the physical presence and support of my friends and family of origin, as well as the comfort and privilege of being there to care for them in times of need. I also let go of the familiarity of the Polish and European sociopolitical systems, along with the financial support provided by the state.

Many women who sought my advice often asked what to do when support is lacking. One of my favorite quotes from C. G. Jung that I share in such moments is: "The highest, most decisive experience is to be alone with one's own self. You must be alone to find out what supports you, when you find that you cannot support yourself. Only this experience can give you an indestructible foundation." I will be sharing more about ways we can support ourselves along the lines.

Being an expat or immigrant provides a unique opportunity to experience this challenging state of mind. Letting go of what is familiar and, to some extent, comfortable is not always easy—it can be surprising and resistant to lasting change, often because "the known" aspects of our personality constitute vital parts of our previous identity. Our "old" *Persona*—a social mask we construct in order to function in family and community, among others—tends to resurface persistently, and many times throughout the day we may find ourselves making deliberate choices to act in ways that support our "new" identity—the one that is still taking shape and remains "under construction."

This process of becoming, introduced by the move abroad, is rarely a smooth journey. It requires a great deal of energy and, in practical terms, relies on maintaining physical and mental well-being. These are themes I often explore with the men and women I work with and encounter daily. However, this process can also lead to identity crises, emotional exhaustion, burnout, and depression—realities that sometimes are a well-kept secret for many expats. Those who dream about moving abroad are often unaware of the complexities and far-reaching consequences of expatriating.

Who has approached the process of identity transformation with full awareness (?) of challenges, such as:

- the unavoidable change in personality and the need to find or create a new identity, or parts of it,
- rebuilding daily life from scratch,
- forging connections and building bridges to new communities,
- relational issues in couples, arising from the fact that both partners change in unpredictable ways, often beyond one's control,

- navigating job searches while grappling with feelings of inadequacy and impostor syndrome,
- coping with loneliness and homesickness,
- dealing with financial constraints,
- facing the lack of support previously available,
- balancing family demands in both the place of relocation and the country of origin.

Living as an expat brings opportunities for adventure and growth, but it also challenges one's sense of self. Identity loss can creep in as the familiar markers of who you are—language, customs, connections—fade in the face of new surroundings. One of the challenges and at the same time seeds of growth lies within learning the ability to befriend this transformative nature of loss. It's an invitation to reevaluate, to redefine yourself beyond cultural labels, and to discover unexpected layers of resilience.

Identity crisis often stems from rejection and also often accompanies expat's life. From feeling excluded from locals to struggling to connect with soul-feeding purpose, career or community can be disheartening. My personal way of viewing rejection as a temporary state, rather than a judgment, softens its sting and allows to see it as a part of a necessary process of growth. What can be a truly nourishing—daily, weekly, or monthly—practice is investing time in finding and reaching out to fellow expat communities that cherish similar values. Such activities immediately foster a sense of belonging. Being consistent in these self-care activities is essential for your mental health as an expat—it brings long-term benefits rooted in a sense of safety and creates meaning by making you a vital part of something larger than yourself. Moreover, like-minded communities create spaces and forums for personal growth by giving you the chance to express your uniqueness and fulfill various emotional and cognitive developmental needs.

Please be mindful that shifts in identity may evoke feelings of vulnerability, anger, or sadness, which can increase your risk of depression. Befriending these feelings begins with letting go of criticism and practicing self-compassion. Creative activities, mindfulness, or therapy can also be very helpful. Living as an expat comes with ups and downs, but every challenge you face helps you grow and build a stronger, more whole, and more integrated version of yourself. This, from a Jungian psychology perspective, is the ultimate goal of every transition.

For me, moving to the USA was an unexpected event, shaped by a series of synchronicities, including saving my younger daughter from a life-threatening accident.

Synchronicity is a phenomenon that appears as coincidence but, in a world dominated by rational, left-brain thinking, challenges the notion that events in our lives occur solely in a linear and chronological way. Some events—such as meeting certain individuals along our journey—happen spontaneously, unplanned, often irrationally, and lead to far-reaching consequences that ripple deeply into the future, impacting not only our lives but also the lives of those around us.

For those who believe life can be planned, I would say that, to some extent and under certain conditions, it is possible. However, the key lies in forging a connection with oneself—intuition, heart, dreams, and the right hemisphere of the brain, which governs creativity, imagination, musical and artistic abilities, spatial awareness, emotional expression, abstract thinking, and more.

How does one plan life abroad, as an expatriate in the USA? In my work with clients, I explore the intricacies of their personal development journeys through bodywork, dreams, symbolic thinking, creating art, and mindfulness practices.

In my case, creativity proved to be one of my most valuable internal resources. It manifested almost immediately as a "helping hand" through meeting an artist, who became my friend and guide through the intricate maze of Californian reality.

Shortly after relocating, I discovered a love for painting, and my passion for creating paintings, illustrations, collages, and photographs began to flourish.

Shortly after relocating there was a moment in my life when I felt the need to express emotions experienced in my new country. While visiting numerous amazing libraries, book stores, watching beautifully illustrated children's picture books, and admiring nature during walks with my children, I discovered a love for creating paintings, illustrations, collages, and photographs. As Jung pointed out, "By not being aware of having a shadow, you declare a part of your personality to be non-existent."

The shadow represents the unconscious aspects of our personality that we tend to reject or deny because they conflict with how we want to see ourselves or how we wish to be seen by others. These could include traits, skills, talents, desires, or emotions that we might consider negative, shameful, or even morally unacceptable. The shadow isn't inherently "bad"—it is simply made up of the parts of ourselves we have yet to fully acknowledge or integrate for the sake of our growth.

As I continued to explore my newfound talent, I found myself drawn to butterflies as a subject. They seemed to embody everything I was experiencing—transformation, growth, and the beauty of becoming. Each painting became a reflection of my own journey, capturing the stages of my life where I had to shed old layers, embrace change, and emerge as someone new. The butterflies symbolized resilience and the delicate yet profound nature of personal evolution. Creating this series of butterfly paintings was more than an artistic exercise—it was deeply healing and empowering. It reminded me that we are always in the process of becoming, and that transformation, though challenging, is a gift. Through painting, I found a way to honor and appreciate my own becoming. I strongly advocate for the use of creativity in consulting practices and wherever else it can be applied. My aim is to explore the artistic process in a creative way that directly allows for drawing meaning and building connections between the art created and the life we live here and now.

My experience confirms that seeking and finding individual meaning in life's events has had profound transformational influence on reclaiming my own power. I would like to highlight that the biggest challenges since coming to the USA for me were holding the tension of opposites, which I always saw as opportunities for the growth of my mindset and personality. Holding the tension of opposites refers to the ability to sit with and reconcile two contradictory feelings, ideas, situations, and parts in ourselves without rushing to resolve the tension between them. It's a concept often discussed in psychology, particularly in the work of Carl Jung, where it involves embracing and accepting conflicting aspects of oneself or one's experience to foster growth and deeper understanding.

I could provide many examples of opposites embedded in the local system and in my personal life. To name just a few paradoxes: being in the heart of Silicon Valley, often surrounded by industrial environments, while cherishing nature and witnessing its gradual encroachment by advancing civilization; living in a comfortable apartment while encountering large communities of homeless people unable to support themselves; or observing the stark disparity between investments in the educational system—from kindergarten to 12th grade—and the significant investments made in corporate or healthcare institutions. As a result of this situation, I discovered and participated in Compassion Cultivation Training conducted by The Center for Compassion and Altruism Research and Education (CCARE)—founded and directed by Dr. James Doty, Clinical Professor of Neurosurgery—at Stanford University School of Medicine in Palo Alto. This training changed my perception of not only expat's hardships, as I could learn deeper about life experiences of Californians, as well as intentionally practice compassion, altruism, and empathy. They are prosocial and positive qualities, innate to our minds, that form the foundation of our common humanity. These qualities serve as a remarkable resource, akin to the concept of the Jungian collective unconscious, enabling us to draw upon our emotional group wisdom—especially during challenging times—helping us not only to endure but to thrive. That is partly why the culture of the San Francisco Bay Area amazed me with its strong need for community building, its remarkable capacity for it, and the ease and joy it brings.

2.4 MEANINGFUL COINCIDENCES SUPPORT OUR INTENTIONS

> And, when you want something, all the universe conspires in helping you to achieve it.
> —Paulo Coelho, The Alchemist

One day, my friend called me to say that a local interpreter agency was seeking a Polish medical interpreter to support patients in hospitals at Stanford in Palo Alto. I regarded this phone call as a synchronicity, as I had been searching for a suitable opportunity that aligned with my skills and family obligations.

Synchronicity, as mentioned earlier, is a concept introduced by Carl Jung to describe meaningful coincidences that occur without a clear cause-and-effect connection, yet seem deeply significant. Unlike random chance, synchronicity refers to instances where events align in a way that feels purposeful or symbolic, often carrying a personal or emotional resonance for the individual experiencing them. For example, you might think about an old friend you haven't spoken to in years, and then receive a message from them shortly afterward. Or perhaps you're grappling with a big decision, and suddenly encounter a book, song, or stranger's remark that provides insight or clarity on the matter. Jung believed these events are expressions of an underlying unity in the universe and can often reflect an alignment between the external world and our individual psyche.

After passing a qualifying exam, I became part of a community of professionals dedicated to saving the lives of children who came to Palo Alto for heart surgeries. My background in active listening, communication, psychology, mediation, and compassion became a powerful toolkit of internal resources, enabling me to support many Polish families seeking help for their children in the USA. Working in this environment taught me invaluable lessons—about cultural differences, determination, and resilience demonstrated by young children, their parents, doctors, and other medical staff—and opened doors to new competencies and decisions.

2.5 FINDING PATH WITHIN

> The meeting of two personalities is like the contact of two chemical substances: if there is any reaction, both are transformed.
>
> —Carl Gustav Jung

> An essential part of creativity is not being afraid to fail.
>
> —Edwin Land

I was combining my role as a medical interpreter with childcare, while my daughters attended a parent-participatory school and preschool. However, the Covid outbreak in February 2020 radically changed our lives. While managing the challenges of schooling rebellious children at home through online devices, I gradually stepped away from my interpreting position and transitioned into becoming a solopreneur in the fields of art and psychological consulting. This shift aligned with my educational background in psychology and my deep interest in the psyche, creativity, and natural problem-solving skills.

I have always admired how children perceive the world. They are a profound source of inspiration to me, because they are masters of the state of flow—effortlessly accessing their creative spark. This resonates with the perspective of actor John Cleese, who famously remarked, "Creativity is not a talent. It is a way of operating." It certainly is. I would go even further to say that we have a fundamental right to creativity—because creativity is not an option, it is an essential part of our mental health, rooted in our connection to the self.

Two of my favorite researchers in the field of psychology are Arnold and Amy Mindell. Arnold Mindell (1987), expanded Jungian concepts and created process-oriented psychology. In our waking life, we may not be fully aware of the role and effect our dreams and physical symptoms really have. Mindell (1987) integrated these aspects into psychological work, creating interesting, insightful tools for inner development stemming from the combination of psychology, physics, and spirituality. Mindell's concept of "deep democracy" allows for every part of personality to be heard, to coexist, and to draw from the diverse, equally legitimate needs each part of our psyche has. Without domination or diminishing of any voice, we can deeply and wisely approach solving internal conflicts, which we tend to project on the outside objects in personal, professional, and political lives alike. This approach is a valuable resource for expat women. It is helpful in navigating cultural challenges encountered abroad, especially in identifying inner experiences—like feelings of isolation, loneliness, identity shifts, adaptation struggles—and dynamics present in their lives.

Attentive exploration of deep processes behind these experiences helps grow self-awareness and psychological resilience. Starting point is the curious, non-judgmental exploration of the present moment. There is a certain amount of energy located in particular in the feelings of stress, anxiety, or depression, which can be released through skillful use of proper psychological tools.

Self-awareness and emotional resilience are fundamental in building authentic, soul-feeding connections, both with oneself and diverse communities. A healthy mind creates growth and balance and contributes to personal and collective well-being on many levels.

2.6 WHAT DO EXPATS AND MONARCH BUTTERFLIES HAVE IN COMMON?

> Every time you are willing to say "Yes" to everything on your path, you express the hero inside of you.
>
> —Maria Nemeth

> If the path before you is clear, you're probably on someone else's path.
>
> —Joseph Campbell

When I was looking for some answers to create meaning from the story of my own migration, an apparent symbol of a monarch butterfly (Danaus plexippus) appeared to me as I passionately painted more than a dozen butterflies during the first two years after moving to the USA. California is a state famous for habitats of that species. Monarchs are known for their never-ending movement from south to north and back—from Argentina through Chile, Brazil, and Peru to Mexico, Florida, and Canada—with a long stop in Santa Cruz, California, an hour from my apartment. They have been living in that cycle for at least two million years … leaving home, going on a great journey of their lives, completing multiple life cycles, and returning to their place of origin. Their hero's quest resembles that of many expats.

What is even more interesting is that "the monarch genome enhances our ability to better understand the genetic and molecular basis of long-distance migration," and "no genetic difference is seen between migrating and nonmigrating monarchs, but the gene is expressed in migrating monarchs and not expressed in nonmigrating monarchs" (Zhan et al. 2011). It seems that migrating monarchs are a super-species with butterfly superpowers and the fastest-evolving insects on Earth.

Monarchs' migration is a wonder rooted in strength and adaptability. It contains hidden symbolic meanings—different for each of us, yet universal. Thanks to the necessity of that cyclical movement, one of the most fascinating migratory species fulfills its life purpose and draws its power from transformation. A journey from caterpillar to butterfly during their distant flight back and forth mirrors my expat's journey, which is about metamorphosing and becoming. I wonder what good comes of it this time?

ACKNOWLEDGMENTS

This chapter came to life thanks to the support of numerous people in my life. Among them is Dr. Manleenjot Kaur, who reached out to me after seeing my involvement in *The Expat Woman*, founded by Nyna Caputi, whom I met while celebrating

International Women's Day in 2023 in Palo Alto, California, during a gathering of *The Polish Professional Women of Silicon Valley.*

I deeply appreciate the hundreds of inspiring conversations I have had since arriving in the USA in 2017 with women, men, and children who are dear to my heart. In particular, I am grateful to Solveiga Jaskunas, Anna Marecka, Barbara Supłat, Krzysztof Wisiński, Aneta and Paweł Czepczor, Alyssa and Alex Abramets, and many more amazing individuals I have encountered in the USA.

I also want to express my gratitude to my dear Polish friends—expats, travelers, and artists—Marzena Zdunek, Magdalena Markiewicz, Dominika Renes, Małgorzata Renes, and Andrzej Renes.

Special appreciation goes to my adventurous and brave husband, Stan, and our two daughters, Antonina and Zofia. Without your presence in my life, this chapter would never have seen the sunlight or starlight.

I am profoundly thankful for my dear father, who instilled in me an everlasting admiration for nature and shares my passion for conversations about life. And to my mother, who passed away in 2021 and will forever live in my heart—she profoundly experienced and shared with me the magic of synchronicity, believed in life's fortunate twists and turns, and found a practical framework for the creative feminine spark during the monumental sociopolitical and economic transition in Poland in 1989.

I draw endless inspiration from the field of depth and process psychology, especially from Carl Jung and his contemporary disciples and followers. Among them are Dr. James Hollis, Marion Wooodman, Dr. Clarissa Pinkola Estes, Lisa Marchiano, Deborah Stewart, Joseph Lee, Dr. Avi Goren-Bar, Arnold Mindell, and Amy Mindell. Their thoughts nourish my soul on my *Heroine's Path* and *Wild Woman Journey*.

Lastly, I want to thank all the individuals within the vibrant, multicultural melting pot that is the international community of the San Francisco Bay Area. You are masters at balancing the tension of opposites that coexist in the unique Californian realm. You create an extraordinary atmosphere while redefining and preserving the meaning of the contemporary American Dream.

BIBLIOGRAPHY

Jung, C. G., Jaffe, A. (1989). *Memories, Dreams, Reflections*. Vintage Books.

Marchiano, L., Stewart, D., Lee, J. (2024). *Dream Wise*. Unlocking meaning of your dreams. Souns True.

Mindell, A. (1987). *The Dreambody in Relationships*. Penguin Books.

Viorst, J. (1986). *Necessary Losses*. Simon & Schuster.

Zhan, S., Merlin, C., Boore, J. L., Reppert, S. M. (2011). The monarch butterfly genome yields insights into long-distance migration. *Cell*, 147(5), 1171–1185. https://doi.org/10.1016/j.cell.2011.09.052

3 Global Grit
Job Adjustment Challenges of the Indian Self-Initiated Expatriates

Manleenjot Kaur, Surbhi Kapur, Sheela Bijlwan and Sonam Bhaduriya

3.1 INTRODUCTION

The Indian expatriate population is one of the most agile around the globe. According to the United Nations, in 2017, around 17 million expatriates were Indians. Some of the prime destinations for Indian expatriates are the USA, Canada, New Zealand and Australia, UAE and Saudi Arabia. While in the Middle East Indians are working as manual laborers, in the USA they are highly skilled professionals in mathematics, science, business, finance and law. Since more and more Indians are travelling to other countries, they do face major cultural differences in the countries they expatriate to. The cultural differences may be the customs, rituals, values and behaviors (Figure 3.1).

Uncertainty Avoidance Index (UAI) is one of the dimensions of Hofstede's cultural model which defines the society's tolerance for uncertainty and ambiguity. Power Distance is a scale that measures the difference between a person in a higher position and one in a lower position. The powerful people have more influence in the society. The more the distance, the more the difference. Masculinity vs. Femininity is the aspect of culture to behave on masculine traits like pay, salary, rewards, promotions, compensations or the feminine traits like goodwill, compassionate work behavior, cooperation among peers. Individualism vs. collectivism connotes that in individualistic cultures, the person works toward his own personal growth and goals, whereas in collectivist cultures, the person is oriented toward the whole group or family to which he belongs. Long-term vs. short-term orientation indicates that the people in long-term–oriented cultures seek long-term relationships with organizations and projects, whereas the people in short-term-oriented cultures believe in short-term employment and negotiations.

Self-Initiated Expatriates (SIEs):

SIEs are people who, on their own volition and initiative, seek and find work overseas (Inkson and Myers, 2003; Lee, 2005). These individuals are distinct from organizational expatriates (OEs), who are sent by their employer to an overseas subsidiary

DOI: 10.1201/9781003604402-3

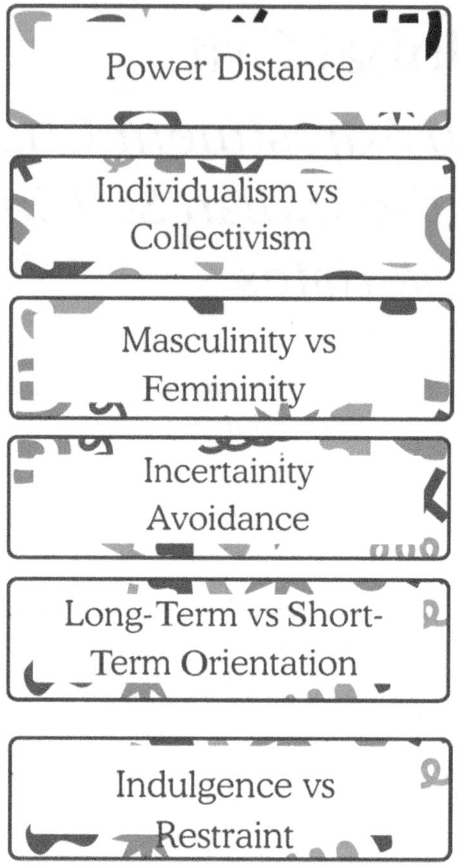

FIGURE 3.1 Hofstede's dimensions.

Work Adjustment:

Theory of work adjustment suggests that there is an interaction between the employees' skills and abilities and the employers' job requirements, and the present study refers to this theory for a better understanding of the equivalence of SIE's work abilities and the requirements of the milieu they work in (Dawis and Lofquist, 1984).

Motives

Expat-Insider 2018 states that the Indian expatriates move abroad to find a job there.

3.2 LITERATURE REVIEW

SIE is a workforce who initiate their career path on their own (Cerdin and Selmer, 2013). According to Doherty et al. (2013), an expatriate is called a SIE when he decides and implements his stay in the country on his own accord. The SIEs can be bifurcated into subcategories: young opportunists – who move abroad for a long time; job seekers – who have better job opportunities and financial considerations as their main motive to move abroad; officials – who move abroad for better job conditions, better pay scale and new experiences; localized professionals; international professionals; and dual career couples (Suutari and Brewster, 2000). There are some hurdles that are only faced by SIEs when compared to OEs like they are jobless, have to find residence and are not backed by any organization or management. There are always some push and pull factors behind one's decisions to move abroad. The push factors in case of SIEs are economic problems and unemployment. And the pull factors add to luring job market, family consideration and financial aspects. The motives of SIEs to move abroad are for career and job opportunities, family considerations, better job opportunities, better pay scales and adventure and travel (Tharenou and Caulfield, 2010; Peiperl et al., 2014; Wechtler, 2015). The previous studies say that even though SIEs start their career earlier than the OEs, they work under supervisors, have less on-the-job embeddedness (still they get lower positions which signifies the underutilization of their skills) and they have lesser organizational support.

3.3 CULTURAL DISTANCE AND CULTURAL DIVERSITY

The cultural distance plays a very important role in adjustment (Ellis, 2012; Peltokorpi, 2008). Hofstede researched on IBM employees in 55 countries to come up with 5 dimensions on which culture of every country can be distinguished from other. Culture diversity plays a very important role at workplace. The people with different cultures contribute to different ideas to the growth of the organization, and the negative effects include communication gap or misinterpretation of information between multicultural people. Every country has its own culture which can be defined as a set of values and language, religious aspects which makes it different from the other one (Amadeo, 2013).

UAI – This indicates that people do not tolerate ambiguity and stick to rules and regulations. The people in cultures with higher UAI feel uncomfortable in uncertain situations. UAI of India is 40, whereas that of the USA is 46, Australia is 51, Singapore is 8.

The research question here is whether UAI is related to the work adjustment of Indian SIEs and the variables are as under:

Variables for UAI:

- Anxiety at work
- Likelihood to continue to work at the present organization
- Uncomfortable with rules and regulations
- Care for the employee's ambitions
- Cultural difference affects their performance

Variables for Work Adjustment:

- Utilizing skills
- Pay

- Promotion
- Social standing
- Autonomy

3.4 OBJECTIVES OF THE STUDY

To explore the motivation of Indian SIEs to move to the USA, Singapore and Australia.

To find out whether SIEs feel discriminated, left out of important work tasks and positions, are anxious at work and less adjusted due to uncertainty and cultural difference (UAI).

To identify the hindrances SIEs face while working abroad.

To suggest measures they need to take to better absorb in the foreign culture.

3.5 METHODOLOGY

The empirical research methodology was used. Questionnaires were sent to 40 SIEs in the USA, Singapore and Australia through Google Docs because the sample was distributed in a larger geographical area. Variables for UAI were taken from (Kapur et al., 2015). Variables for Work Adjustment were taken from Job adjustment inventory by Zainol et al. (2014). The questionnaire was prepared on the basis of these two. The questionnaire consisted of 40 questions out of which 10 questions were demographic.

Reliability Statistics	
Cronbach's Alpha	**No of Items**
0.911	37

Three questions were open ended. The Cronbach Alpha for the questionnaire was 0.91. In this study, the correlation method was used for the variables of UAI and work adjustment to find whether the better adjustment of Indians at workplace is due to lower UAI score.

3.5.1 Key Findings

Individual Characteristics	
Gender	
Females	18
Males	22

Age	
55 and above	2
45–55	5
35–45	17
25–35	16
18–25	1
Marital Status	
Married	25
Unmarried	12
Divorced	3
Have Children	
Yes	25
No	15
Respondents' Countries	
USA	15
Australia	15
Singapore	10
Living Abroad Since	
5 years and above	13
4–5 Years	3
3-4 Years	6
2–3 Years	2
1–2 Years	1
Less than 1 year	16
Education	
Doctorate	1
Masters	19
Graduation	19
High school	2
Variables for Work Adjustment	
Utilizing skills	
Autonomy	
Pay	
Promotion	
Social standing	

Working in the Same Organisation	
Since 5 years and above	22
Since 3–5 years	4
Since 1–3 years	4
Since less than 1 year	11

Motives	
Career opportunities	14
Regional considerations	1
Job conditions	4
Family considerations	15
Financial considerations	4
Adventure and travel	2

Variables for UAI
Anxiety at work
Persistence with the same organization
Comfortability by rules and regulations
Care for the employee's ambitions
Cultural difference affects their performance

Variables for Work Adjustment
Utilising skills
Autonomy
Pay
Promotion
Social standing

3.6 RESULTS

3.6.1 UAI Variables

Anxiety was not a problem at workplace for 69.23% respondents. 41% respondents wanted to continue their job with the present organization. 74.35% respondents were not uncomfortable with the rules and regulations of the organization. 59% respondents believed that their ambitions are taken care of in the organization. 79.48% respondents stated that the culture had no negative effect on their performance at the workplace.

3.6.2 Variables for Work Adjustment

77% respondents believed that their abilities matched the skills that were required in their daily tasks. 69% respondents did not feel discriminated at their workplace, and 49% respondents had decision-making roles in their organization. 77% respondents

TABLE 3.1
Correlation Table

	Utilising Skills	Pay	Promotion	Autonomy	Social Standing
No anxiety	0.334	0.503	0.682	0.382	0.382
Job continuation	0.318	0.491	0.677	0.363	0.363
Rules and regulation comfortability	0.520	0.577	0.447	0.705	0.705
Ambitions taken care of	0.627	0.643	0.797	0.513	0.513
Cultural difference acceptance	0.494	0.451	0.412	0.799	0.799

did not feel lonely and isolated at the workplace which meant that they had good social standing in the organization. 92.3% respondents had autonomy at their workplace and their supervisors trusted them for their important tasks.

Promotion is highly correlated with whether ambitions of the employee are taken care of (0.797) which means that if ambitions are well taken care of, then the promotion chances of the person are quite high. Cultural difference acceptance is highly correlated with autonomy at work (0.799) and social standing of the employee (0.799), as shown in Table 3.1.

The study also revealed that 64% respondents felt that no measures were taken to motivate the expatriates at their workplace which can help them through the cultural barriers. Moreover, 84.6% respondents marked that no multicultural training was being given to them. This reveals that despite no cultural training, Indians are well adjusted in foreign countries with lesser UAI. Even 79.5% respondents believe that if they were sent through an organization, there would be no difference in their cultural amalgamation. The SIEs believe that their path of coming abroad on their own opened new opportunities for their career and also for their social networks. They shared that the path of self- initiation was difficult and long, but the new challenges helped them gather a lot of information about policies, rules and regulations of the country which helped them in long run.

The following major hindrances were listed by the respondents:

- It takes time to understand cultural differences since one mostly handles an Asia Pacific role.
- Not being open to the new culture surely leads to acute feeling of hindrances and as also not trying to understand the host culture and comparing it with one's own.
- Language problem.
- Cultural difference, likes, dislikes with regard to nationals.
- Breaking host country culture (customer-centric, self-controlled).
- Assuming intent also creates problems in building communication with host nationals.

- Accustoming to their culture and their cultural norms can be a difficult task.
- Systematic racism exists at managerial levels, which may impact acquiring promotions (for non-native people).
- Lack of local experience which some companies ask for when one first arrives in a foreign country.
- Absence of language fluency may result in communication barriers
- Politics of the country affects the work environment as also racism, stereotypes, discriminatory attitudes and behaviors, bullying, gossiping etc.

3.7 MEASURES

The measures suggested and being taken by Indian SIEs to get better adjusted in foreign countries were as follows:

- Understanding the culture, specifically the work culture, as having acceptance, inclusivity and equality around the workplace in a new culture makes the transition process easier for people who have recently migrated from other countries.
- Being open minded and respecting differences of opinion can help furthermore.
- Participating in social events and social activities at work and more socializing with the locals and not being restricted to only same nationality mingling.
- Respecting every person and their job/role, no matter how big or small.
- Being knowledgeable about one's work and exhibiting confidence in one's ability.
- Being more open to ideas.
- Working on language skills.
- Realizing that at the end one is dealing with people and people are professional and emotional both.
- Commencing one's career with volunteering work to understand the environment, as host country organizations usually prefer expatriates with experience and may not provide a chance to prove their abilities.

3.7.1 At the Organizational End

- All employees to be treated alike and multicultural employees to be respected and valued for their contributions at work.
- Everyone's opinion should matter and the workplace to be progressive enough to adapt to all cultures.
- Celebrate other ethnic group's festivals with equal enjoyment and try to respect as that would have a deeper meaning and impact than just providing a day off to the Indian SIEs.

And finally,

- Training and constant education along with cultural get-together make it the right mix to adjustment.

3.8 CONCLUSION

The SIEs are the people who start their own path of expatriation without any support from the organization or any other agency. The majority of the SIEs are younger in age and highly qualified. The major motives to move abroad found in this study are family considerations and career opportunities. SIEs seem to have no difficulty in work culture in countries like the USA, Australia and Singapore. They face lesser difficulty in socializing with the host people. There has been observed no lack of job autonomy as against what has been mentioned in previous research (Lee, 2005). The earlier studies (Oluwadare, 2019) stated that SIEs were discriminated, had lesser salary, but this study contradicts the statement as the respondents wrote that they were not discriminated at their workplace but were supported by their peers and supervisors. The studies also depicted (Peltokorpi & Froese, 2009) that SIEs had lesser job satisfaction as their skills were underutilized which is again not seen in this study. UAI is related to the work adjustment. If the UAI of the countries is same (host and parent), the adjustment is better in the workplace, as is revealed in this study. Singapore has a UAI of 8 but the difference in work adjustment as compared to the USA and Australia was not seen due to small sample size but a larger sample size could throw more light on this aspect.

The training or any initiative is not vigorously noted by SIEs to make them comfortable by the host countries, but as UAI is less for India, the Indians get easily adapted to the foreign countries with lesser UAIs. The hindrances as voiced by SIEs were majorly cultural differences. Though they adapt themselves to the different cultural environment, it is a major roadblock to easy amalgamation in a foreign land. The SIEs use social networking as a major tool to overcome culture shock. Language or communication is another hindrance for SIEs, but as they have good off-the-job embeddedness (Meuer et al., 2019), they overcome this too. Lack of local experience in the beginning is the real challenge, along with less knowledge of the rules and regulations of the country. The Indian SIEs should respect other cultures to get their culture understood. As the UAI is same in countries like the USA and Australia, the adjustment issues are usually with culture and language which can be easily handled with some measures taken.

3.9 RESEARCH LIMITATIONS AND RECOMMENDATIONS

This study finds out the real challenges faced by the SIEs in foreign countries and also at the workplace. We found out the measures they are taking to adjust better abroad. The sample size is very small due to the time constraint which resulted in almost same responses from across the three countries. However, it will help as a pilot study for the future research. Only three countries were taken in this study because of the sample collection convenience. In future research, the sample can be taken from other countries with higher UAI than India. Only the SIEs were taken as sample because there is a lack of research in this particular population. Besides SIEs, OEs can also be studied in future research.

REFERENCES

Amadeo, K. (2013). Cultural diversity in the workplace. The Balance. http://useconomy.about.com/od/suppl1/g/Cultural-Diversity.htm

Cerdin, Jean-Luc & Selmer, Jan. (2013). Who is a self-initiated expatriate? Towards conceptual clarity of a common notion. *International Journal of Human Resource Management*. 25. https://doi.org/10.1080/09585192.2013.863793.

Dawis, R. V. & Lofquist, L. H. (1984). *A Psychological Theory of Work Adjustment: An Individual-Differences Model and Its Applications*. Minneapolis: University of Minnesota Press.

Doherty, Noeleen, Richardson, Julia & Thorn, Kaye. (2013). Self-initiated expatriation. *Career Development International*. 18. 97–112. https://doi.org/10.1108/13620431311305971.

Ellis, D. R. (2012). Exploring cultural dimensions as predictors of performance management preferences: the case of self-initiating expatriate New Zealanders in Belgium. *The International Journal of Human Resource Management*, 23(10), 2087–2107.

Inkson, K., & Myers, B. A. (2003). "The big OE": Self-directed travel and career development. *Career Development International*, 8(4), 170–181.

Kapur, Surbhi, Hotta, Ashish & Mahato, Ashok. (2015). Asian Journal of Management Research Online Open Access publishing platform for Management Research Cracking the cross-cultural code: Barriers and strategies to successful survival in foreign organizations.

Lee, J. (2005). Effects of leadership and leader-member exchange on commitment. *Leadership and Organization Development Journal*. 26(8). 655–672. https://doi.org/10.1108/01437730510633728

Meuer, Johannes, Tröster, Christian, Angstmann, Michèle, Backes-Gellner, Uschi & Pull, Kerstin. (2019). Embeddedness and the repatriation intention of assigned and self-initiated expatriates. *European Management Journal*. 37. https://doi.org/10.1016/j.emj.2019.03.002

Oluwadare, Sunday. (2019). Culture shock and job expectations adjustment among Nigerian self-initiated expatriates in the United Kingdom. *International Journal of Human Resource Studies*. 9. 81. https://doi.org/10.5296/ijhrs.v9i1.13865

Peiperl, M., Levy, O., & Sorell, M. (2014). Cross-border mobility of self-initiated and organizational expatriates: Evidence from large-scale data on work histories. *International Studies of Management & Organization*, 44(3), 44–65.

Peltokorpi, Vesa. (2008). Cross-cultural adjustment of expatriates in Japan. *International Journal of Human Resource Management*. 19. 1588–1606. https://doi.org/10.1080/09585190802294903

Peltokorpi, Vesa & Froese, Fabian. (2009). Organizational expatriates and self-initiated expatriates: Who adjust better to work and life in Japan? *The International Journal of Human Resource Management*. 20. 1096–1112. https://doi.org/10.1080/09585190902850299

Suutari, Vesa & Brewster, Chris. (2000). Making their own way: International experience through self-initiated foreign assignments. *Journal of World Business*. 35. 417–436. https://doi.org/10.1016/S1090-9516(00)00046-8

Tharenou, P. & Caulfield, N. (2010). Will I stay or will I go? Explaining repatriation by self-initiated expatriates. *Academy of Management Journal*. 53(5). 1009–1028. https://doi.org/10.5465/AMJ.2010.54533183.

Wechtler, H. (2015). Cross-cultural adjustment of female self-initiated expatriates: A longitudinal diary study. In *Academy of Management Proceedings* (Vol. 2015, No. 1, p. 13414). Briarcliff Manor, NY: Academy of Management.

Zainol, Halmi, Wahid, A.M.A., Ahmad, A.C., Tharim, Abdul, Hamimi, Asniza & Ismail, Ain. (2014). The influence of work adjustment of Malaysian expatriate executives in Malaysian construction companies overseas. *E3S Web of Conferences*. 3. 01020. https://doi.org/10.1051/e3sconf/20140301020.

4 Repat Diaries
Experience Working as a Repat in the USA

Amy Amann

4.1 INTRODUCTION

The phenomenon of professional repatriation – returning to one's home country after extended international work experience – represents an increasingly significant but underexplored aspect of global career mobility. While extensive literature addresses expatriation challenges and cross-cultural adjustment for those moving abroad (Black et al., 1991; Ward et al., 2001), substantially less academic attention has been devoted to the reverse journey: the complex process of professional and personal reintegration that occurs when global professionals return home.

This gap in scholarly attention is particularly noteworthy given the growth in international career mobility and a common assumption that "coming home" should be straightforward and simple. Recent research suggests that repatriation presents unique psychological and professional challenges distinct from initial expatriation (Szkudlarek, 2010), yet personal narratives and detailed case studies remain scarce in the academic literature. The limited available research often focuses on corporate-sponsored expatriates returning to predetermined roles, leaving unexplored the experiences of independent global professionals navigating repatriation without organizational support structures.

This autoethnographic case study addresses a void in the literature by providing a detailed first-person account of an author's professional repatriation in the context of an American returning to the USA after over two decades of international career experience. This is accomplished by referring to established theoretical frameworks such as the W Curve Theory of cultural adjustment (Gullahorn & Gullahorn, 1963), Hofstede's cultural dimensions theory (2001), and contemporary research on cultural intelligence (Earley & Ang, 2003) to contextualize my personal observations within a broader academic understanding of cross-cultural transitions.

The research questions guiding my exploration include: How do established theories of cultural adjustment apply to my contemporary repatriation experience? What specific workplace challenges did I encounter in professional reintegration? How could I leverage my international experience as an organizational asset during repatriation? What factors facilitated or complicated my readjustment process?

DOI: 10.1201/9781003604402-4

Methodologically, I use an autoethnographic approach that combines personal narrative with cultural analysis. This provides both insight into my lived experience of repatriation as well as a broader cultural commentary on American workplace dynamics from my perspective as a returning global professional. This approach is particularly suited to repatriation research, as it captures the subtle dynamics between the cultural and personal experience that quantitative studies might miss, while providing the analytical framework necessary for academic contribution.

The context of my particular repatriation during the COVID-19 pandemic and, independent of a supporting corporation or other organization, the associated changes in work patterns, social interactions, and economic conditions add a contemporary relevance. It also highlights how environmental disruptions in personal and professional contexts can compound traditional repatriation challenges. These aspects provided a unique lens to examine adaptation processes under conditions of heightened uncertainty and rapid change.

I hope my study contributes to repatriation literature by providing a contemporary account that bridges academic theory with practical experience, offering insights valuable for scholars studying global career mobility and practitioners supporting repatriation processes. Through critical examination of my workplace reintegration, cultural readjustment, and professional identity reconstruction, I illuminate the complex, multi-dimensional nature of professional repatriation while demonstrating how international experience, despite presenting readjustment challenges, creates valuable competencies that can benefit both individuals and organizations in increasingly globalized work environments.

4.2 CROSS-BORDER CAREER TRAJECTORY

Numerous sources concerning topics of international relocation and settlement exist among people who could receive the tags of “expat,” “immigrant,” “digital nomad,” or other designations. And of course, there is a literature on the repatriation journey that also touches upon general issues such as the reverse culture shock and the fact that the home environment has changed. However, my investigation, both official and unofficial, has led me to very few sources, where individuals spread their personal accounts and experiences conspicuously and candidly, in the process of repatriation after a long time away, in the motherland. I hope that this autoethnographic account can bridge theory and reality by demonstrating how established cultural adjustment theories uphold for professionals, particularly for those navigating transitions without organizational support.

Foundational works like “The Art of Coming Home” by Craig Storti address general repatriation challenges. While these provide essential conceptual frameworks for returnees, recent academic research has increasingly focused on the career implications of international assignments and the complexity of professional reintegration within evolving organizational contexts (Baruch et al., 2016). I seek to build upon existing literature by providing a contemporary perspective from someone who functioned essentially as an independent global professional, emigrating voluntarily and later choosing to repatriate without corporate relocation support, compounded by the additional complexities of the COVID-19 pandemic.

4.3 CULTURAL IMMERSION AND LEADERSHIP ABROAD

My experience was fairly unusual in terms of how and why I was living abroad, which likely shapes my perspective on independent repatriation. I wasn't technically an expatriate seconded by my company for a limited-duration contract. I didn't relocate to be together in a relationship, nor did I have family or relatives to lean on. I hadn't studied abroad and remained in the country after the studies. And I certainly wasn't a retiree, entertainer, or free-lance digital nomad working remotely while traveling the world.

What I was: an American who secured employment in another country – initially Germany. I lived in one location and secured traditional employment on a local contract. As my career gained momentum over time and new career prospects emerged in various countries, I moved to those new countries. In that regard, I essentially emigrated from the USA and immigrated to Germany, then subsequently to the Netherlands, back to Germany, with a project secondment to South Korea (where I was the expatriate), and finally to Luxembourg for another corporate position. From there, I returned to the USA, thus completing what cultural adjustment literature terms the full cycle of expatriation and repatriation (Gullahorn & Gullahorn, 1963).

This reading offers valuable background to the interpretation of my own views on independent repatriation, both in the literal and the figurative sense of that word. I had relocated multiple times, consistently maintaining the status of foreigner, developing what Ward et al. (2001) identify as enhanced cultural adaptation competencies through repeated cross-cultural transitions. This pattern of serial international mobility created familiarity with culture shock cycles and the professional reinvention processes that characterize what Baruch et al. (2016) term "global career patterns."

I navigated these transitions independently, working with the hiring company to secure permits for residence and work authorization. Sometimes I was the only foreign employee. Sometimes I was the only American. Sometimes both. Some positions I accepted were newly created, meaning there was no structured onboarding, no cultural orientation training, and no inclusion preparation provided to teammates on how to collaborate with their new colleagues. Paradoxically, it was often my responsibility as an organizational development specialist to design and develop these programs; I used my own experience to inform the content and structure of new hire onboarding and training.

These experiences proved invaluable when I repatriated to the USA, during the COVID-19 pandemic when remote work had become normalized. Vaccination programs were just being deployed, and businesses and offices were just starting to reopen. I recognized that making the move during this unprecedented time was an exception to the typical mobility program. I searched extensively for guidance on "independent repatriation during a global pandemic" but found no relevant resources. I knew that the journey would be a unique one without a pre-scripted playbook.

However, I realized that I possessed several characteristics that would enable the transition and provide guidance as I navigated through it. My background as a qualified coach and leadership trainer provided me with essential resources. I had extensive experience supporting business leaders and professionals, and global teams, and individuals navigating organizational change, professional challenges, and personal transitions. I relied heavily on my coaching techniques to navigate both the physical relocation and entry into the new professional environment. What I was experiencing aligns with what researchers conceptualize as the "W-curve" of cultural adjustment, an extension of the original U-curve that incorporates the repatriation journey (Gullahorn & Gullahorn, 1963). While expatriation involves the familiar cycle of honeymoon phase, culture shock, and eventual adjustment, repatriation triggers a parallel process: initial excitement about returning home, followed by reverse culture shock, and eventually achieving new equilibrium that integrates both international and domestic experiences.

4.4 RETURNING HOME: THE REPATRIATION TRANSITION

This is where my repatriation story truly begins. I entered the Technology and SaaS sector, positioning myself within STEM fields where I could leverage my previous experience in medical and engineering companies. This continuity represents what cultural adjustment research identifies as a form of professional anchoring that entails maintaining expertise domains while navigating cultural transitions (Black et al., 1991).

When I started my new position, we were still deep in the COVID-19 pandemic period. We weren't "back to normal," which itself was being redefined as "the new normal." Many companies, including mine, were operating under hybrid arrangements with two to three days in-office and remote work for the remainder. This contrasted sharply with my experience in Europe, where most companies remained entirely remote. European organizations had generally maintained more conservative return-to-office policies.

Being physically present in the office proved invaluable for my reintegration. It allowed me to observe workplace dynamics and organizational culture firsthand, which Earley and Ang (2003) frame as essential for developing cultural intelligence, including the ability to read social cues, interpret unspoken norms, and adjust behavioral responses accordingly. What I was experiencing aligns with what repatriation literature describes as the need for cultural re-learning, even within one's home country (Szkudlarek, 2010).

The hybrid model created an interesting dynamic for my adjustment process. Office days provided structured social interaction and cultural immersion, while remote days offered processing time to reflect on these observations and recalibrate my responses. Ironically, the pandemic timing worked in my favor: since everyone was adapting to new workplace patterns, I wasn't the only one figuring out cultural norms and professional expectations. Everyone was in a re-learning mode. This collective adjustment period somewhat masked my own repatriation challenges within the broader organizational transition.

4.5 RE-ENTERING THE US WORKFORCE

I knew that the typical American work culture would differ significantly from what I had experienced over the prior 20 years. Having worked mostly in western Europe in international companies, with extensive collaboration across offices in Asia, I anticipated that this multicultural variety and diversity would be absent from the more monolithic US Midwest culture I now found myself in. This shift from multicultural to monocultural work environments represents what cultural research identifies as a significant adjustment challenge for repatriates, often more psychologically demanding than the original expatriation process (Szkudlarek, 2010).

Although I had worked in the context of German multinational corporations, or any other company with a strong parent culture, the prevailing national culture did affect our work strategies, nonetheless there seemed to be some sort of a work culture that played an organizational level: a global work culture that developed into a meta-level organization culture. Our own rhythms and patterns of work evolved and did not depend on any single cultural construct. Relationship-building followed many directions and scripts. Power worked in subtle ways that were nearly invisible and hard to articulate or pinpoint. Cultural intelligence research might characterize this as high-context cultural navigation (Hall, 1976). In our own way, with so many nationalities convening together, navigating cultural differences, whether communicating in English or other languages, reading non-verbal cues, or establishing overtly agreed-upon "ways of working together," was an ingrained part of the workday; employees were aware of this and were motivated to succeed in a diverse environment.

The American work culture I experienced struck me as less nuanced in comparison. I was so accustomed to being sensitive to workplace dynamics and starting a role, trying to identify these to be able to adjust as soon as possible. I was also accustomed to going into a new role with a clear identity as "the foreigner." When I took on the new role in the USA, I was able to shed these habits. However, in a few years, it turned out that this state of affairs was also characterized by its own peculiarities. The types of relationships to negotiate through were decreased, the communication styles were more conventional, and definitely no tricky time zone collocation!

There were a couple of issues that made these experiences different from my international work in the past. As with many times the contrast upon returning to a new culture is subtle, so too was returning to the expected result. They weren't as obvious as many commonly cited differences like limited vacation policies or employment-based healthcare, or employment-at-will. Rather, between the pace and rhythm of work and the implicit norms were differences that surprised me, as an American in reverse migration.

4.6 WORKPLACE READJUSTMENT IN THE USA

Even in small companies within the tech sector, I was surprised at how frequently the boss maintained first and final decision-making authority. This dynamic reflects what Hofstede's cultural dimensions theory characterizes as higher power distance, where hierarchy and authority are more readily accepted in workplace relationships (Hofstede, 2001). Research suggests this pattern may be more pronounced in domestic versus internationally oriented organizations (Baruch et al., 2016). While I understood that executives were accountable for outcomes, I noticed the frequency with which directives were issued to teams, including toward the other senior leaders and subject matter experts.

Taking a cultural lens to this observation, this represents what Meyer (1991) categorizes as "direct negative feedback" culture. Americans communicate decisions explicitly without extensive context or relationship-building preamble. However, some directives came wrapped in softer variations, questions, or suggestions, demonstrating what Meyer terms "upgraders" and "downgraders" that modify directness while maintaining clear hierarchical intent. The message's ultimate intent remained "do as I say."

I came from organizational cultures where substantive debate with colleagues, including supervisors, was welcomed, especially in western and northern Europe. These organizations tend to employ and empower employees as domain experts in their functional areas. Their voices are expected to be heard, and in fact, speaking up about issues becomes a professional responsibility. Even for issues tangential to one's area, employees are charged to think critically, plan scenarios, and consider cross-functional implications. Companies prioritize well-planned operations, based on the premise that relevant expertise yields superior outcomes, making employee input not just welcomed as a collaborative device, but essential to the growth and success of the organization.

The pace of work was notably faster, with communications becoming shorter and messages ping-ponging rapidly between colleagues. While people expressed frustration with meetings, some relied heavily on instant messaging to conduct full discussions. We worked intensively toward resolutions and deadlines, pressing forward despite communication challenges.

Work often shifted direction or scope unexpectedly. Projects pivoted quickly, exacerbating the need for frequent, rapid communication while limiting opportunities to catch up with colleagues or even process developments individually. The quick communication patterns and shortcuts taken, including speculation and the tendency to "run with it" even with partial information, sometimes created confusion. Then projects would screech to a halt, requiring us to finally gather and piece together individual components to understand the situation holistically.

I recalled a cross-cultural training about project management differences between Germans and Americans, where the distinction was drawn that Americans "start, then plan," whereas Germans "plan, then start." This observation aligns with cultural research on uncertainty avoidance and temporal orientation, where American workplace culture typically emphasizes immediate action and quick results over sufficient planning phases (Hofstede, 2001). When I first heard this years ago, I was skeptical. I didn't believe that the American approach was so modest. Over time, I observed it in my interactions with American offices, and now, back in the USA, I was fully confronted with this cultural reality. The speed and cadence were faster, with greater focus on results and less emphasis on process. There was considerable trial and error.

That said, the work quality remained professional, and in the traditional spirit of American resourcefulness, we collaborated effectively to fix the most egregious errors and move initiatives forward. I found colleagues to be friendly, eager to experiment, quick to offer opinions, and skilled at rapidly building rapport to tackle challenges. There was typically good team spirit, and I believe the informal, almost casual nature of workplace relationships facilitated this dynamic. However, there were notable differences in relationship depth and personal connection. Even within the casual work atmosphere, people seemed more focused on professional tasks and tended to maintain boundaries between work and personal lives. They were friendly at work, but maintained what I'd characterize as an invisible fence around authentic personal sharing. Conversations avoided topics like age, finances, or personal struggles beyond socially acceptable complaints about being busy, tax rates, or favorite sports teams. This reflects what Meyer's Culture Map characterizes as "task-based trust" cultures, where professional credibility is built through competence and reliability rather than personal relationship development, a significant shift from the "relationship-based trust" environments I had navigated in European contexts (Meyer, 1991).

The tech companies I joined had relatively young, early-career teams. Colleagues were interested in exploring career development while building experience. Companies were willing to hire at the junior experience level, compensate accordingly, and bring people on to "see how they work out." This quick "trial-and-error" recruitment strategy differed significantly from the careful, structured hiring processes with extensive vetting and candidate assessment I had experienced previously.

Bringing in new hires with limited industry experience typically meant minimal structured training or onboarding. This was an area I could immediately contribute to professionalizing. The fast pace and shifting priorities presented challenges for creating structured programs. Training managers to provide useful instructions and effective feedback to new hires represented another challenge. I had developed appreciation for structured onboarding processes to support new team members. However, people's willingness to experiment, especially with solutions that might ease their workload, helped garner support for new initiatives. The team enjoyed collaborating on ideas, contributing tips and techniques, and someone even offered to create instructional videos. In hindsight, this exemplified the innovativeness and willingness to experiment that research identifies as characteristic features of American organizational culture (Shane, 1993).

4.7 REVERSE CULTURE SHOCK AND CAREER REALIGNMENT

Cognitively, I understood that entering new environments presents challenges filled with unknowns. Returning to a familiar place after extended absence creates similar challenges, though what I was experiencing aligned precisely with what Gullahorn and Gullahorn (1963) conceptualize as the W-curve of cultural adjustment. The repatriation literature suggests people tend to underestimate these adjustments or the gaps between perception and reality that emerge over time, often assuming that "coming home" should be straightforward and linear. The longer I was back in the USA and the more I immersed myself in social settings and work, the more the differences became apparent.

The landscape was filled with subtle, almost invisible ways that made me feel foreign in my own country. From small and seemingly trivial matters to more substantive issues, being confronted with the reality of learning or re-learning became a constant companion during my repatriation process. Research on workplace reintegration confirms that repatriates often experience unexpected challenges in familiar environments due to changed perspectives and evolved professional practices developed during extended international careers (Szkudlarek, 2010). What the W-curve theory predicts – and what I was experiencing – is that reverse culture shock often manifests differently than initial culture shock, frequently through subtle disconnects rather than obvious cultural misunderstandings.

While I was learning the nuances of American work culture, my colleagues were learning to understand the perspective I brought. I had to remember that they hadn't previously worked with a fellow American who had extensive international work experience. I was new to them not only as a coworker but as a construct. At least to them, I was a complete functioning adult who happened to have some serious professional experience, was working a job like anybody else, and therefore had no responsibilities in her culture, or should responsibly do anything about it. However, I also feel that I was a curiosity to be explored and studied as they evolved a functional relationship with me.

Responses between colleagues usually varied between curious and fascinated, occasionally even jealous, and occasionally even suspicious. Most people expressed interest in learning about international work experiences and wanted to hear my story. "Where did you live?" was a common question. When I mentioned having spent many years in Germany, responses often landed on something like "Oh, I went to Oktoberfest once. They have really good beer!" or "My dad was stationed in Germany for a couple years." These are attempts to find common ground while demonstrating their discomfort with being confronted with a fellow citizen with a different career trajectory.

When people realized that my international experience had been focused on professional work rather than traveling to exotic places as a tourist, their idealized notions were quickly deflated. What seemed to them like a lucky or privileged circumstance was, in truth, demanding professional work. When they learned that my life abroad also focused on normal activities like working, paying bills, and pursuing hobbies, I got a sense that their own self-imposed visions of a spectacular Instagram-fed life were erased.

What became clear is that few people in one's home country truly understand or appreciate the professional and personal transformation stemming from extended international careers. This represents what both reverse culture shock literature and W-curve theory identify as a fundamental challenge of repatriation. Their knowledge about the countries I had lived in was based on extremely limited exposure and perspective. For some, these places were unknown or possibly only a spot on a map.

Recent academic research confirms that international career mobility creates unique professional competencies and identity shifts that distinguish repatriates from both domestic workers and temporary expatriates (Baruch et al., 2016). However, the adjustment responsibility falls to the professional in transition to adapt to local context and operate within the established cultural frameworks. While some fortunate circumstances were involved, the international experience had fundamentally resulted from deliberate choices, hard work, persistence, and accepting significant risks.

When it came to actual work contributions, however, the team expected the same behavior and working style. As someone from the same country, I was viewed as "one of the gang." This was comforting to be "an insider"; however, it didn't account for the ways my international experience had changed how I approached work. It created a most fascinating professional duality. This dynamic worked both ways. Integration was probably easier because I had been socialized in American culture, but my colleagues also noticed my different approach to projects, or they reacted unexpectedly to my methods.

These differences were often very subtle and initially difficult to recognize. Over time, with reflection, I realized that different gestures, eye contact patterns, or expressions would land differently with colleagues. People sensed something different based on subconscious cues that made me seem somehow "off," though they couldn't identify exactly what felt unfamiliar. This aligns with both W-curve theory's predictions about reverse culture shock manifestations and research findings that repatriation challenges often appear as subtle interpersonal disconnects rather than obvious cultural misunderstandings (Gullahorn & Gullahorn, 1963; Gaw, 2000).

This observation isn't a judgment of my colleagues but rather an acknowledgment of how humans rely on familiar social cues and interactions to navigate their professional worlds. People naturally relate to and gravitate toward what feels known and familiar. It's the evolutionary wiring that keeps humans safe by treating unfamiliar patterns as potentially threatening until proven otherwise.

Some interactions I remember included forgetting English words for concepts, or thinking of phrases in other languages that better match the topic. Sometimes I pronounce foreign words using native pronunciation. Honestly, there are moments when English words simply escape me randomly and without warning. I recall forgetting terms like "computer keyboard" or "notepad." In the Midwest, where many people have German heritage and family names, I could only pronounce them using German phonetics with long "u" vowels and crisp "t" endings. People's reactions ranged from curiosity to occasional appreciation, but more often they hesitated as they realized I drew on different cultural references.

I found myself handing out business cards with two hands and a slight bow, as I had become accustomed to in Asia. I shook hands with people I met, following European colleague practices – hugging, or bear-hugging, isn't common. I missed having champagne with colleague's birthday cakes. I preferred a small glass of water on my desk versus large water bottles. I turned off the monitors to conserve electricity, and took the stairs.

While these stories seem amusing in retrospect, they created confusion for both sides in the moment. I encountered apps, services, and products I hadn't used before. There were numerous mobile applications I'd never heard of. Much of Europe hadn't adopted app-based solutions for every transaction, so using my phone for booking or paying for services wasn't a habit. European data privacy norms and reluctance to share personal information for routine transactions contrasted sharply with American digital payment expectations. I didn't have Venmo or Zelle accounts, nor did I understand their functions. Repaying a colleague for lunch became a ten-minute ordeal to download an app and complete registration. Cash was no longer readily accepted. My colleagues showed bemused patience when I asked about Venmo's functionality, graciously explaining its features and demonstrating how to use it.

Since my recent prior roles had been in finance and professional services, transitioning to tech companies represented additional adaptation challenges. These were various software applications that I heard of and not much used. I was flung into modern digital processes with shared documents, cloud drives, and collaborative programs. Although I enjoyed lots of things, it was certainly an accelerated learning and catching-up process.

I learned how to conduct my professional duties. A relocation had not reformed the core competencies, but there were new tools and systems. It turned out to be an educational process with severe ascents and descents, necessitating regular team inquiries, unavoidable errors, and undergoing course corrections almost instantly. I encountered tools and processes which were not new to me, and these embody their own specialized vocabulary that needed to be adapted as with new linguistic systems. I needed to walk the fine line between unplanned scenarios and trying to develop systematic solutions. This was not the sole challenge that came with repatriation, but it was endemic to growing organizations in the tech industry.

Regarding day-to-day processes, besides the superficial modulation which was quite easy to handle, the overall process ran without major problems. Team life was much the same as other times but slightly different. Formal order was abandoned in favor of so-called informal daily meetings. Dashboard reviews began to be used as weekly reports. What had been framed until then as brainstorming was now laser-oriented on goal-setting and on getting measurable results. Together, each of these changes produced no major effect, but as a whole, they defined new work rhythms and intensity. A gradual low-level compulsion to perform, keep visible, be active, contribute vocally, and demonstrate effort. This certainly was a pervasive approach to ensuring people remained active and connected to both their work and team. It encouraged individual contribution and advancement while maintaining checks and balances to ensure collective progress.

At this point, I was experiencing the W-curve's final phase: gradual adjustment and integration. Rather than simply returning to my pre-expatriation professional identity, I was developing what might be characterized as a hybrid professional approach that incorporated international perspectives within American organizational contexts. This aligns with the W-curve theory's prediction that successful repatriation creates a new equilibrium rather than restoration of previous states (Gullahorn & Gullahorn, 1963).

Even with my own personal integration, it didn't fail me to observe gender roles in American workplaces. I sensed that there are more women in senior leadership positions generally in the USA, and within my technology and innovation company experiences, both genders seemed to contribute relatively equally. Fortunately, I haven't encountered or had to address gender-based harassment issues in my HR or coaching roles.

However, glass ceiling dynamics still exist in certain contexts. Marketing roles, customer success positions, and some scrum master or business intelligence positions were typically filled by women. Technical roles, coding positions, design, strategy, and architecture positions were predominantly male. Perhaps my perspective is skewed by Midwest regional patterns that don't represent broader national markets. But traditional role divisions remained evident in many contexts. I mention this not to critique gender gap issues but rather to provide awareness for newcomers and returnees. While there are significantly more freedoms and opportunities, with increased women's representation in senior organizational levels and entrepreneurial spaces, we haven't achieved 50–50 balance, and role distributions remain inconsistent at best.

4.8 WHAT HELPED IN THE JOB: THE HIDDEN ASSETS OF INTERNATIONAL EXPERIENCE

It's all the new workplace with new rhythm, new habits, and here and there a hiccup or a blooper. These are natural and expected elements when into a new professional environment. I experienced all of these, whether expected or not. And, I managed to adapt and make my way. What was less obvious at least at first, yet ultimately more valuable, was recognizing how my accumulated international experience had fundamentally transformed my professional capabilities.

Repatriation research establishes that returning home after international assignments requires different adjustment cycles and methods, sufficiently different from domestic relocation, or from the initial expatriate assignment (Black et al., 1991). Yet as recent comprehensive reviews demonstrate, what we know about leveraging these experiences continues to evolve (Chiang et al., 2018).

The reality became clear: my years of professional experience in large, well-run international organizations didn't diminish when crossing time zones or borders. Instead, these experiences create a wealth of knowledge and insights, as well as personal and interpersonal capabilities that prove invaluable to the modern workplace and business demands. This represents perhaps the most crucial message for repatriates and new immigrants: we carry an arsenal of transferable competencies that transcend geographic boundaries.

4.8.1 Cultural Intelligence

International experience develops what researchers term "cultural intelligence," the capability to function effectively in diverse settings (Earley & Ang, 2003). This goes far beyond simple awareness; it's the adaptability skills that become valuable organizational assets (Bonache et al., 2001). Current research confirms that these competencies represent measurable outcomes with significant implications, which help position international experience as a strategic advantage rather than merely personal enrichment that is "cool" or "interesting." What is indeed interesting is that working successfully in another culture arms professionals with both soft skills and essential business capabilities that companies hire for, particularly those focused on innovation, growth, and transformation.

Living abroad cultivates a broader worldview and develops sophisticated information processing capabilities. I became accustomed to absorbing disparate information and rapidly making sense of complexity. Over time, I developed new thinking patterns and systems to handle information, like an enhanced method for intake, analysis, and dissemination of information. This skill sharpens critical thinking abilities which are applicable across many situations and contexts.

The corporate competence of "managing ambiguity" takes on new meaning after we've navigated Byzantine bureaucracies, negotiated in foreign languages, or found our way through unfamiliar cities at midnight. These experiences create what organizational behavior research identifies as transferable competencies – skills developed in one cultural context that enhance performance in seemingly unrelated situations (Spencer & Spencer, 1993).

4.8.2 New Application for Skills

International professionals develop mastery in self-directed learning by necessity. When systems, habits, and processes that assumedly "everyone knows" remain mysterious, success depends on rapid knowledge acquisition and gap identification. This autonomy becomes a competitive advantage in any professional setting.

Paradoxically, what research identifies as "reverse culture shock" can contribute to enhanced professional capabilities. For me, the adjustment process (Chiang et al., 2018) forced me to approach familiar environments with analytical skills developed abroad. Even while it was challenging to navigate the new environment, the abilities I had honed abroad were the ones that enabled me to complete the navigation.

The additional effort required to prove oneself, such as for perceived gaps in local educational credentials ("go to the right school"), language fluency, or cultural knowledge, particularly pop culture, consumer trends, and politics, paradoxically creates a secret weapon: the ability to see with fresh eyes. Approaching systems afresh exposes gaps and inefficiencies that familiarity hides. Challenging established processes uncovers better alternatives, and digging deeper often surfaces insights overlooked by established teams.

International experience generally requires extraordinary organization. Having to live two lives and stay in touch with the past and create a new life through balancing between various profiles, tasks, and administrative demands. The same scenario creates more applicable workplace skills: deadline management, proactive communications, awareness of stakeholders, and systematic attention to detail.

4.8.3 Resilience and Continuous Improvement

Trustworthiness, tenacity, and accountability were other positive impacts/items that I acquired and had during the international experience. A failure is used to create something to be feared into information that spurs growth. Post-mortems, deconstruction, incorporation of feedback, and identifying ways of improvement are all second nature after a while as we learn when our previous experience goes wrong and get ready to approach the next experience.

This is in line with corporate competency of continuous learning and development, but not as an aspirational purpose, but as a life practice that is perfected by need. Resourcefulness, or creativity and adaptability, only makes them sharper as they hone their skills as their life continually throws them trial and error without provisions.

4.8.4 Practical Application: Scaling and Transforming Business

Once I got my new position, it was really new with no definite definition or performance expectations. It became important to rely on practical experience. I needed to analyze the needs of the organization, identify the trends, and find suitable solutions. I referred to systems and methods of past organizational structures relevant to successful growing companies.

Two decades in business came in handy in understanding the problems with scaling, and the necessity of low-cost, efficient solutions to get through stressful periods of growth. With no obligation to adopt specific approaches or practices used in the comparable firms, I could cooperate with the team contributing to creating practice solutions.

Success required courage to take initiative, push beyond personal comfort zones, and engage with diverse stakeholders without letting perceived barriers impede progress. With a limited local network of former colleagues and resources, every step required research and proactive outreach. It was akin to starting over in a new country: stepping outside comfort zones, asking essential questions, and finding solutions. These had become fundamental skills that became an integral part of my playbook.

These qualities enabled me to bridge gaps and contribute unique perspectives that proved essential. The international experience that initially seemed like a potential disadvantage revealed itself as a comprehensive toolkit for professional excellence.

Contemporary repatriation research validates this transformation, confirming that the adjustment process, while challenging, generates measurable individual and organizational outcomes (Black et al., 1991). The key lies in recognizing these capabilities as strategic assets rather than merely personal experiences, or worse, an obstacle to overcome. In sum, the international experience is a foundation and accelerator for professional performance.

4.8.5 Recommendations for Repatriates: Lessons from the Field

My experience represents one report among countless others in the repatriation landscape. While individual journeys remain unique, the coaching work I conduct with other returnees reveals consistent patterns and practical strategies that prove helpful regardless of specific circumstances. Many people return at some point during their international careers, often without extensive planning or realistic preparation for what lies ahead.

This guidance emerges from both personal experience and professional coaching practice. One can always find something to apply it to and some piece of knowledge that can relate to certain circumstances or struggles that returnees struggle with in their course of adjustment.

4.8.6 Accepting the Reality: There Is No "Back"

The simple stereotype is that repatriation means going back to what is familiar. No such destination exists. A workplace structure is evolving extraordinarily fast: throughout 2020–2025 alone, a professional environment was changing in response to home work area, massive layoffs, seemingly big quit, moving silently, a silent quit, and more than RGQ. The conditions the returnees will face in the workplace are only vaguely comparable to the ones they left.

More importantly, overseas experience is a period of far-reaching individual change. Significant disruptions in our lives such as prolonged cultural immersion produce profound changes, much beyond surface assimilation. Although fundamental personality characteristics are preserved, outlooks change radically, the senses become extended, and ways of work are developed under the pressure of the environment.

The returnee brings with him radically different information processing and problem-solving as well as understanding working relationship dynamics. Studies confirm what effective repatriates learn over time: integration yields more favorable results than either extreme assimilation upon returning to the native workplace environment or isolationism.

As the acculturation framework established by Berry (1997) illustrates, long-term adjustment is more beneficial when valued elements of international experience are retained, while being able to adjust to the workplace norms of the local place. Instead of grieving about declines in prior professional identity or seeking to recreate the conditions of the past, the thriving repatriated can look forward to the rediscovery of another world and themselves in it. This perspective shift transforms potential loss into opportunity for continued growth and development.

4.8.7 Preparing for Professional Immigration

The transformational process includes entering the realm of professional immigrant treatment – itself thus a differentiated professional category for therapists of whatever status – whether citizen or not, regardless of whether they have had professional employment in the domestic private sphere. Warm organizational welcome (already working there before) is rare. Without sustained close ties and continuously renewed internal networks, returnees return in a de facto "brand new" capacity and need systematic reintegration.

This reality can be brutal – especially for those that expect praise or appreciation for travel abroad. However, by understanding this dynamic, you can formulate realistic preparation and strategy moving forward. Companies and colleagues haven't kept room for returning multi-professionals, and don't necessarily know how to weave international experience productively. In addition to professional and observable differences, returnees experience what we call "ambient disconnections," which are behaviors that someone can feel are not right, but where they cannot clearly determine what it is that seems disordered.

These disconnects manifest in surprising colleague responses, communication misunderstandings, or social interaction patterns that don't fit with what was learned while on international assignments. The challenge is then either to explore and resolve those disconnects, or to live with them provisionally until a pattern can be identified clear enough to name and adjust. Thus, experience gained in the international ground begins to pay dividend in the form of tolerance to ambiguity, which is the required condition for the successful reintegration into the criteria of the home.

4.8.8 Systematic Network Rebuilding

Intellectual geographies change constantly on moving from country to country and on employment, so that complete network rebuilding rather than brief reconnection with former acquaintances must take place. Industry groups change, companies emerge and go out of business, key people move on to different roles, and trusted partnerships fall apart.

Professional infrastructure – which provided stability – might not even exist in recognizable forms anymore. Systemically engaging in network rebuilding involves revamping knowledge of developments in the community, recognizing the key players currently in play, understanding new groupings, and acknowledging the presence or even saliency of formerly used resources.

This networking process closely reflects those activated when gaining access to a completely different professional market, and are skills developed and transferable by international professionals into the domestic workplace. The work involves a departure from comfort zones in meeting new contacts and contributing to professional networks that had found form in the absence. Something that may have reliably existed in the past no longer exists, in which case new solutions have to be found.

4.8.9 Reframing Visibility and Contribution

Many returnees find it uncomfortable when at work; they either ask their views or want them to be part of the conversation. This pain is usually caused by fear of not being able to share their global experience or feeling like they will stick out as an oddball and be different from team members. The reframe is a change in thinking that requests for input indicates the value of different points of view from their colleague.

Different cultures offer differing points of view that will enhance performance in teams as well as organizational decision-making. Without this kind of information, teams lose out on key resources that emerge as they learn from one another cross-culturally. Generally speaking, that contribution level will rise in concert with greater comfort levels and more team cohesion. Instead of downplaying experience overseas, successful repatriates learn to strategically share their experiences as well as build credibility and reflect an ability to work collaboratively with others.

4.8.10 Maintaining Learning Orientation

Continual learning is hereby considered a precondition for a successful adjustment of repatriation. This means being watchful, attentive, and purposely analytical. Flexibility skills that result from international experience are valuable underpinnings to domestic re-adjustment issues. Reinvention doesn't just happen to particularly brave persons. It's open to everyone in good health who's prepared to take first steps (and who is persistent enough to carry on through the adjustment process).

Evidenced ability in successfully adapting to previous international assignments shows ability to handle current challenges domestically. The learner-orientation is extended from technical competence knowledge, through recognition of social cues, pattern modification of communication, and reacquaintance with cultural norms. Virtually all of these adjustments are subconscious and therefore harder to find and correct than gross deficiencies in a student's skills.

4.8.11 Managing Comfort and Growth

As a result, one must understand that the comfort one wishes from the "house" world is based on familiarity and certainty. These qualities take time to recover from. While it can be frustrating when old habits are hard to resume, the discovery and development of a new behavior can be extremely rewarding. Growth is non-negotiable at all ages and phases of our careers. Stagnation makes no contribution to individual development nor organization's contribution.

Good international experience has shown what transmutations and adaptations can occur. Repatriation is another means of holistic change or evolution rather than a retrograde back to earlier conditions. This may include reskilling, upskilling, or finding new ways for capabilities to be shared, which have been built up from experience across countries. Engaging in domestic professional practices and the formation of practices that enable integration, to assimilate in a different culture, remain important. Addressing parochial concerns with sticks of domestic identity and useful edible balls of global identity is an ever-present and growing task.

4.8.12 Timeline Reality and Patience

Extensions of time are always longer than expected, especially for long-term expatriates. A reasonable timescale for effective integration is a year or longer. Consistent with the literature, repatriation adjustment stages are often reported to be longer than the period of early expatriation adjustment, with the added burden of expectancies of rapid reintegration adding to psychological difficulty (Ward et al., 2001).

Hurrying through the process or creating artificial deadlines will be counterproductive. Each member of a group develops at different speeds in different areas of adjustment – professional practice, social adjustment, identity development, and environmental reintroduction. Cultivating a perspective that embraces natural rhythms, a baseline view that welcomes small steps, lowers stress upon oneself and leaves space for natural growth. The longer timeline is mediated by adjustment complexity instead of disenchantment or exogeny. Most long-term dictators have time deficits for substantial integration into all dimensions of life.

4.8.13 Celebrating Distinctiveness and Strategic Positioning

Given the vastness of the territory for growth, fitting institutions into former boundaries for the sake of uniformity is unsustainable and unprofitable, whereas allowing specific character guarantees a more solid basis for growth. I know that international experience can produce durable differences of perspective, approach, and capability that cannot be readily overcome through domestic experience alone. Signs of success are easier interactions on a day-to-day basis, starting some new relationships, and becoming more comfortable in a professional role.

Each milestone deserves to be celebrated and congratulated because it represents a successful trajectory of adjustment action. Self-advocacy is a skill that cannot be learned without rehearsal, especially in environments where there is no risk and the stakes are relatively low. Achievements are achievements regardless of experience. It has simply happened in different geographical places. The key is to speak plainly about capabilities and accomplishments with no disclaiming phrases or geographic modifiers.

As you plan and strategize around the hierarchical or chain structure of the organization through values of contribution, it is critical to consider how individual contribution value impacts that positioning. Online challenges support team e-learning for building competencies that increase team performance and capabilities for the organization. These contributions, when understood, can help overcome imposter syndrome while generating confidence for career growth.

4.8.14 Approaching the Journey with Realistic Optimism

Repatriation can be successful if you enter the process with full commitment and understanding that preparation can only take you part of the way. Many developments have not been foreseen or cannot be planned for before they occur. The first is the basis of being as good as one can be, given information while trusting that experience does come with purpose that will become apparent later. This perspective involves balancing expectations that are grounded in reality with optimism about future results.

Nothing is ever easy, and expats always face challenges they didn't know or think of, while finding options out of range of their imagination. The development of adaptability in response to international experience provides an excellent basis for handling whatever conditions arise. Travel requires remembering to be patient with oneself, persistence in times of difficulty, and trust in developed adaptive capacities.

While the conditions differ significantly for each person, the basic potential for adjusting adequately to a new environment is present in everyone who has successfully completed international professional assignments. When applied austerely and systematically, the skills that rendered the country internationally successful are the ones to facilitate reintegration domestically.

4.9 CONCLUSION

My analysis of professional repatriation to the USA demonstrates the multi-layered quality of "coming home" after years of international professional experience. Finally, and in another attempt to broaden the relatively limited and anachronistic literature on repatriation by sharing a contemporary and first-person account of repatriation which gives voice to too many of us, this book discusses in depth how I am reacclimating, culturally re-adjusting, and re-constructing my work identity.

My research is consistent with existing theoretical models and also demonstrates nuanced aspects of repatriation which require deeper focus in future research. The W-lesson of culture shock (Gullahorn & Gullahorn, 1963) has been especially true to me, as initial excitement of being home contributed to subtle but continual reverse culture shock and then slow reintegration while still holding a unique blend of my international and domestic work identities. However, in my experience, the timescale for this adjustment is much longer than expected. My full reintroduction was at least three years long, much longer than my initial adjustment to expatriation was.

I found the process of workplace reintegration to be a rather complex one, comprised of three interrelated, but distinct factors: a structural adjustment to different organizational cultures, an interpersonal adjustment to single culture ideas about teams, and techniques for adapting to evolved systems and processes (technology). Szkudlarek (2010) has argued that repatriation includes special psychological and technical issues related to initial expatriation, while expanding the understanding of how these problems emerge in modern work environments that are technology based.

My experience illustrates a paradox at the core of the repatriation process: while I struggled with severe readjustment challenges, my international experience equipped me with organizational capital in the form of cultural intelligence (Earley & Ang, 2003), increased adaptability, and advanced problem-solving skills. So for organizations that can identify and activate these competencies, potential reintegration challenges can turn into competitive advantages in a more innovative environment.

Several implications emerge from my experience for both future repatriates and organizational practitioners. First, repatriation should be conceptualized as a distinct career transition requiring structured support and attention, not merely a "return to normal." Second, the integration approach to acculturation (Berry, 1997) – maintaining valued international perspectives while adapting to local norms – proved most effective for both my personal adjustment and organizational contribution. Third, timeline expectations must be realistic, acknowledging that comprehensive reintegration is a multi-year process requiring patience and intentional effort.

I acknowledge that my case study has limitations, including its singular perspective and specific geographic and industry context. Future research might explore repatriation experiences across diverse industries, geographic regions, and cultural backgrounds to develop more comprehensive theoretical frameworks. Additionally, longitudinal studies tracking independent repatriates over extended periods could provide deeper insights into the long-term trajectory of professional reintegration.

The COVID-19 pandemic context of my repatriation adds another layer of complexity that warrants further investigation, as remote work normalization and social distancing may have both facilitated and complicated my reintegration process in ways that will influence future repatriation patterns.

Ultimately, my autoethnographic account demonstrates that successful repatriation requires viewing the experience not as a return to a previous state, but as a forward-moving process of professional and personal integration. My international experience fundamentally changed me, creating new competencies and perspectives that, when effectively integrated, have contributed significantly to both my personal fulfillment and organizational contribution. For the growing population of global professionals, I believe understanding repatriation as a complex but navigable transition, rather than a simple homecoming, is essential for career success and personal well-being in our increasingly interconnected world.

REFERENCES

Baruch, Y., Altman, Y., & Tung, R.L. (2016). Career mobility in a global era: Advances in managing expatriation and repatriation. *Academy of Management Annals*, *10*(1), 841–889.

Berry, J.W. (1997). Immigration, acculturation, and adaptation. *Applied Psychology*, *46*(1), 5–34.

Black, J.S., Mendenhall, M., & Oddou, G. (1991). Toward a comprehensive model of international adjustment: An integration of multiple theoretical perspectives. *The Academy of Management Review*, *16*(2), 291–317.

Bonache, J., Brewster, C., & Suutari, V. (2001). Expatriation: A developing research agenda. *Thunderbird International Business Review*, *43*(1), 3–20.

Chiang, F. F., van Esch, E., Birtch, T. A., & Shaffer, M. A. (2018). Repatriation: what do we know and where do we go from here. *The International Journal of Human Resource Management*, *29*(1), 188–226.

Earley, P.C., & Ang, S. (2003). *Cultural Intelligence: Individual Interactions Across Cultures*. Stanford University Press.

Gaw, K.F. (2000). Reverse culture shock in students returning from overseas. *International Journal of Intercultural Relations*, *24*(1), 83–104.

Gullahorn, J. T., & Gullahorn, J. E. (1963). An extension of the U-curve hypothesis. *Journal of Social Issues*, *19*(3), 33–47.

Hall, E. T. (1976). *Beyond Culture*. New York: Anchor Books.

Hofstede, G. (2001). *Culture's Consequences: Comparing Values, Behaviors, Institutions and Organizations Across Nations* (2nd ed.). Sage Publications.

Meyer, H. H. (1991). A solution to the performance appraisal feedback enigma. *Academy of Management Perspectives*, *5*(1), 68–76.

Shane, S. (1993). Cultural influences on national rates of innovation. *Journal of Business Venturing*, *8*(1), 59–73.

Spencer, L.M., & Spencer, S.M. (1993). *Competence at Work: Models for Superior Performance*. John Wiley & Sons.

Szkudlarek, B. (2010a). Reentry—A review of the literature. *International Journal of Intercultural Relations*, *34*(1), 1–21.

Ward, C., Bochner, S., & Furnham, A. (2001). *The Psychology of Culture Shock* (2nd ed.). Routledge.

5 Being Resilient and Financially Fervent in USA

A Personal Narrative from Sandra to You

Sandra Morno

5.1 INTRODUCTION—GET TO KNOW THE AUTHOR

I am Sandra Morno,

I would like to share a personal perspective—woman to woman. This is not the polished biography often seen on LinkedIn, but rather the honest journey that led me to where I am today—helping women like you retain more of your earnings while pursuing your dreams internationally. The reality is, I never initially planned to work in taxation. It wasn't even part of my career aspirations. My goal was to become a sports agent, negotiating deals, advocating for clients, and helping them make sound financial decisions. I even attended graduate school with this purpose in mind. I was confident that this was my destined path. However, life had different plans. During my studies, I needed to support myself financially. Bills do not pause for dreams, so I took an evening job at a friend's accounting firm. What was supposed to be temporary ended up being more than that. I found the work to be monotonous and unfulfilling; I didn't want to become "the tax lady." Initially, I did the work out of necessity, feeling like I was settling. I believed I was on the way to becoming a sports agent, which was my true passion. It took approximately three humbling years and encouragement from a friend and mentor before I recognized what was truly in front of me. Once I embraced the reality, I realized that taxes are much more than numbers—they involve helping individuals recover their rightful money from the Internal Revenue Service (IRS). Many people seek tax assistance because it can be a source of stress, fear, and uncertainty. They want to ensure everything is handled correctly and to have someone advocating on their behalf. I witnessed clients walk in confused, and I was able to help them understand the process with confidence. Showing them strategies to save money transformed their outlook. That was the moment I developed a genuine passion for this work. Furthermore, I discovered a sense of freedom in this profession. I could set my own hours, build a career around my life, be there for my loved ones, travel, learn, and grow. I did not set out to be a tax strategist; rather,

 DOI: 10.1201/9781003604402-5

someone recognized my potential and showed me the meaningful impact I could make—especially for women like us. Women managing businesses, relocating across countries, nurturing families, building dreams, and confronting societal expectations of difficulty and complexity. Although I never planned to become the "tax lady," embracing this unexpected path turned out to be one of the best decisions I've ever made. Today, I empower women like you to retain more of your money, reduce tax-related stress, and pursue your dreams freely across borders. Life may surprise you, but I've learned to trust it—more often than not, it guides you exactly where you need to be. Keep dreaming, keep moving forward, and most importantly, remain unapologetically true to yourself.

5.2 MONEY AND TAXES

Let me provide some context regarding my background. I am the eldest daughter of a first-generation Haitian American family. Our journey was marked by sacrifice, hard work, and resourcefulness rather than wealth or privilege. My parents emigrated here seeking better opportunities, and in our household, the importance of financial responsibility was clearly emphasized: work diligently, be grateful for what you have, and avoid wastefulness.

We didn't talk about investments or taxes or building wealth. No one sat me down and said, "Here's how credit works," or "This is how you make your money grow."

Money was something you *earned* with sweat and determination. It was for rent, food, and bills. Maybe a treat here and there. And the biggest fear? Not having enough and making things stretch. We didn't have the luxury of playing around with money. There was no cushion, no safety net. The lessons were all about survival. And I carried those lessons deeply up into my 30s. I learned to respect money. To hustle for it. To feel grateful any time I had "enough." But what I didn't learn was how taxes really worked. Or how to use the system to my advantage. For us, the IRS was like the legal Mafia. You pay up and they decide what you get back. Tax time meant *stress*. My parents did their best and followed what their accountants told them. My dad was a cab driver making cash income and my mom was a Nurse. I grew up with this idea that taxes were just this thing you hoped wouldn't bite you too badly. You paid what they told you to pay. Crossed your fingers. Praying it was enough. Taxes were this once-a-year stress-fest. Something you dreaded. Something you hoped wouldn't surprise you too badly. The funniest thing is, as I was working as a tax professional, I got my $2500 happy gram from the IRS. You would think I knew better, but I didn't.

At the time in 2013, that amount felt enormous. I remember opening the letter and feeling my stomach drop. The panic was real.

I can still remember asking myself:

What did I do wrong?
How did this happen?
Why didn't anyone tell me?
What if I can't pay this? What happens now?

It wasn't until I spoke with my friend/boss at the accounting firm that he explained why it happened. I had this painful realization:

I could have avoided this.
If I had known better.
If someone had taught me.
It was humbling. Frustrating. But it was also *empowering*.
After going through that situation, I got serious about *understanding* taxes.

That day I made a promise to myself:

If I ever owed money again, it would be because I *planned* to owe it.
I didn't want to feel blindsided again.
And honestly? I didn't want *any other woman*, especially other daughters of immigrants, first-gen kids, hustling entrepreneurs, to feel that way either.
Because we work too hard.
We sacrifice too much.

That IRS bill hurt. But it was also the moment everything changed for me. I needed to stay in this industry to help others.

It taught me the value of *knowing the rules*.
I want you to *keep* what you've earned.
I want you to feel *safe* and *smart* about your money.
I want taxes to be one less thing holding you back from the life you want.

And that promise? It's still what drives me today.

Here's the truth: Most of us didn't grow up talking about taxes or how to keep more of our money. I sure didn't! But getting blindsided by the IRS taught me a huge lesson: knowledge is your best protection.

5.3 MY MOTIVATION FOR TRAVEL AND ITS IMPACT ON MY PERSONAL GROWTH

Travel has always been something I *loved*.

Growing up, we cruised a lot as a family. I lived in several states. Even as a kid, I was curious about new places, new people, new ways of living.

But in 2018, after almost five years in Houston, I woke up one morning and felt this heavy weight in my chest.

I was *tired*.

Tired of the rat race.

Tired of working just to pay bills.

Tired of watching life happen while I stayed in the same place.

I wanted more out of life.

I didn't want to look back one day and wonder, *Is this really it?*

So, I did what so many of us do when we're hungry for change: I started researching.

I asked myself, *Where could I go? Where could I actually live safely as a Black woman?*

Thailand kept coming up, but it was so FAR!! But it would allow me to live without worrying about the hustle. I thought I had it all mapped out in my head.

But life had other ideas.

God had other plans.

In July 2018, I had to move back to Miami for health reasons. It was eye-opening, humbling, and frustrating. It felt like I was pressing pause on everything I wanted. But that time also saved my life. My body was telling me something and I had no choice but to listen.

The dream didn't die, it was delayed. As time progressed, I got stronger and more in tune with my health.

Finally, in 2020, I decided to test myself.

I booked my first **solo flight** to Morocco.

It wasn't glamorous. It was scary as hell. I questioned myself the entire time: *Can I even do this alone?*

When the day came, I had a complete meltdown at the airport. I was like what am I doing! Am I nuts to be trying this?

But there was an airline desk agent, an angel in disguise, who saw me breaking down and helped calm me. She didn't have to, but she did.

Because of her, I got on that plane. My nerves were still there, but I got on that plane not knowing what to expect when I got to Morrocco.

Once I was there, I stayed with a host family which was cool. I did my volunteer work and during downtime wandered the streets of Morocco, tasting the food, meeting people, and giving a little bit of French I could when people spoke to me.

That trip made me realize *I could do this.*

I *did* do it!

I came back at the end of January 2020 so proud of myself. Ready. Determined. Telling myself: *This is it. You're going to travel for real.*

And then! the world shut down! What was supposed to be a month ended up lasting over a year.

Sadly, the onset of the COVID-19 pandemic resulted in prolonged travel restrictions, grounding my plans for over a year. While this was a setback, it reaffirmed my desire to live more fully and authentically, rather than adhering to a safe but limited routine.

In January 2021, inspired by a friend's trip to Mexico, I joined her journey, which further expanded my perspective. Since then, I have traveled and lived in countries including Argentina, Peru, Panama, Greece, Italy, various regions of Mexico, and Colombia. These experiences taught me to slow down, cherish simple pleasures like people-watching at cafés, savor street foods without rushing, and trust my instincts when lost.

Travel has demonstrated that life can take many forms; it's okay to change course, start anew, and pursue different dreams. It showed me that success does not require exhausting oneself or conforming to external expectations. As a Black woman, these journeys have highlighted both the challenges and the empowering beauty of embracing my identity.

Ultimately, travel rekindled my connection with myself. It reminded me that joy, well-being, and adventure are not just luxuries—they are fundamental rights. When asked why I travel, I share that it grants me freedom, peace, and a deeper awareness of who I am. My goal is to inspire other women to realize that even small trips can profoundly influence our lives. We deserve to prioritize ourselves, take that first step, and let the adventure unfold on our terms.

Travel transformed not only my surroundings but also my mindset. It helped me slow down, breathe, and reclaim my life according to my values. I hope every woman can experience that same liberation and empowerment—you don't have to wait for the perfect moment. Start where you are, embrace the journey, and live fully at your own pace.

5.4 UNDERSTANDING THE CHALLENGES AND DEFINING MY PURPOSE

Living abroad has taught me that the curated highlight reels shared on social media often do not reflect the full reality.

Travel can be inspiring, transformative, and freeing.

However, it can also come with significant challenges.

There is a misconception that living overseas means you have everything figured out—that you are lucky, carefree, and living the ideal life.

But I know the truth behind those pictures. I know what it feels like to sit in a place, looking at the mountains from my apartment, that could make you cry, and still feel completely *alone*. I know the worry that hits you at 2 a.m. in a new country, wondering if I've made a terrible mistake. I know the anxiety of not knowing if renting a place, then finding out the place is unsafe. That felt so scary. Having to pivot quickly without panicking. I know the moments of wanting to go home but not even feeling like a stranger at "home."

I've met so many women on this path, women who are brave enough to want something different, bold enough to get on that plane, determined enough to figure it out as they go. We all shared that tiredness together which made the bond we created so much stronger.

Trying to make a living from our laptops while the Wi-Fi cuts out. Buying e-sims with the hope it will work.

Worrying about visas. Feeling like I am supposed to be grateful all the time even when struggling.

And for women? For black and brown women? There's another layer people don't talk about enough.

Being *hyper-aware* of where it's safe.

The pressure to prove I belong.

The exhaustion of being the "first" or the "only."

Feeling eyes on me in places that don't see many people who look like me.

When I talk to women like this, I don't see someone who's lost.

I see someone who's *fighting*.

Fighting to build the life she actually wants.

Fighting to write her own story.

Fighting to say *yes* to herself in a world that taught her to always put herself last.

And that? That's why I'm here.

Because I've felt all of it.

Because I *am* her.

I understand the desire for more than merely existing—to live fully, boldly, and authentically. I aspire to explore the world on my own terms, to challenge external narratives, while still seeking safety, peace, joy, and freedom. To feel confident that my decisions are sound, allowing me to move forward without fear.

My mission is not to promote an idealized "digital nomad" lifestyle but to acknowledge its realities—the joys and the fears, the adventures and the need for careful planning, the beauty and the challenges. For me, it's about affirming that it's okay to want this and that it's worth pursuing. It's also about reminding women that they do not have to navigate this journey alone. Women can embrace both freedom and stability, adventure and safety, responsibility and dreams. We have the power to define our own way of living.

For me, this is more than work—it is personal. It's a reaffirmation to my younger self, who once felt stuck in Houston, staring at the ceiling at 3 a.m., yearning for more but unsure how to achieve it. It's a promise to the woman nervously departing for Morocco, determined despite her fears.

It's my promise to every woman who has ever said, *I want more*.

Because I want you to know it's okay to want more.

It's okay to change your mind.

It's okay to choose *yourself*.

This mission is about holding space for that.

For the questions. For the fears. For the plans. For the dreams.

It's about sitting with you and saying, *I see you. I get it. Let's figure it out together*.

Because you don't have to have it all mapped out. Just getting started is saying a lot!

You just have to want it enough to try.

I want to remind you that:

You're not crazy for wanting this. You're not selfish. You're not naïve. You're *brave*. You're *capable*. You're *allowed*.

You're exactly the kind of woman who can make this life her own. And guess what, let's say you try and it's not for you. That's okay. The point is you know you gave it your all and have no regrets. And you can start again if you choose. This isn't about having it all figured out. It's been about choosing *yourself* even when it's hard. I'm here to remind you: it's okay to want more. You deserve that and then some. You're not too late. You're not asking for too much. You just have to be brave enough to want something different. And that's more than enough.

5.5 WHAT I DO TODAY AND WHY IT MATTERS

These days when I look at my life, I see something I never would have pictured years ago.

Not because I didn't want it.

But because I didn't even know it was an option.

Growing up, I learned to do what was expected: work hard, don't waste, be responsible. That was survival. That was respect.

And I'm grateful for that. It taught me resilience. It taught me to keep going when things got hard.

But no one told me it was okay to want more than just getting by.

I had to figure that part out for myself.

I had to learn that you're allowed to change your mind about the life you want and I'm *still* learning that to this day!

That you can be thankful for what you have and still want something different. (Also still learning … LOL.)

That you can love where you're from and still want to see what else is out there.

That realization didn't come easy.

It took redefining what success meant to me not what the world told me it should mean.

It took moving back home when my health demanded it.

It took letting go of plans and dreams I had mapped out so carefully.

It took giving myself permission to accept the "failures" that were really just redirections.

And along the way, I met so many women who were doing the same. That, in itself, was refreshing.

Women who left jobs, homes, and relationships because something inside them whispered, *"This isn't it there's more."*

Women who packed up their lives to move somewhere new not because it was easy, but because they *had* to find a life that felt like theirs.

That's one of the biggest reasons I do what I do.

Because I want them to know they're not asking for too much.

That wanting to wake up excited is possible.

That wanting to see the world is okay even if people say it isn't "safe" or "realistic."

That wanting to live differently doesn't make you ungrateful. It makes you *honest*.

And here's where taxes come in …

Most people don't connect taxes to freedom, but *I do*.

Because when you're an American living abroad, taxes don't go away. In fact, they get more complicated.

You still have to file every year.

You still have to report your income even if it's not from the U.S.A.

You still have to worry about things like the FEIE, foreign tax credits, FBARs, exchange rates, and what country you're physically in (and for how long).

It's a lot.

And no one hands you a guide when you move overseas.

I've seen so many women feel overwhelmed and frustrated, thinking,

"Am I doing this right?"

"Did I miss a form?"

"Am I going to get a letter from the IRS while I'm 5,000 miles away from home?"

That stress is real.

And it can take the joy out of what should be a beautiful, freeing experience.

That's why I do what I do.

Because when you understand how your money works, how to structure your business, how to legally reduce what you owe, how to plan instead of panic, you give yourself *options*.

You stop overpaying. You stop feeling anxious. You stop holding your breath during tax season.

And suddenly?

You can afford to take that trip, stay abroad longer, or slow down without sacrificing everything you've built.

I blend tax strategy with travel, not because it's trendy but because it's *transformational*.

Because I've seen what happens when women realize they don't have to choose between freedom and financial security.

That's when things really start to shift.

That's when they give themselves permission to live differently.

And that's what I want more women to feel.

Because I've been there.

I've doubted myself.

I've stayed up wondering if I was being irresponsible for wanting *more*.

But I also know the peace that comes when you realize … this is *your* life.

You get to write the rules.

You get to choose your next chapter.

I want women to have that moment, that deep exhale where they think:

"I'm allowed to want this."

We deserve quiet and still mornings.

We deserve to laugh as we navigate languages we're still learning. And watch people's faces look at us like what is she saying.

We deserve to wander and to wonder.

And most of all?

We deserve to know we're not alone.

When I met other women traveling like I was, it brought me peace, comfort, and connection I didn't even know I needed.

Now, I want to be that for someone else.

This isn't just about travel.

It's not just about taxes.

It's about building a life that feels true to *you*.

And that's something *every* woman deserves.

At the end of the day, I'm not just helping women file taxes—I'm helping them reclaim their time, their peace, and their power.

I'm helping them build lives they can be proud of—lives that feel *free*, *grounded*, and deeply *theirs*.

Because when you finally understand how your money works especially as an American living abroad you stop living in fear.

You stop second-guessing every decision.

You start moving with clarity.

You start living with intention.

This shift allows them to embrace new ways of living, to give themselves permission to redefine their lives on their terms. I want women to experience that moment of relief—the deep exhale that says, "I'm allowed to want this."

We all deserve moments of quiet, mornings of reflection, and the joy of navigating new languages with humor and curiosity—still learning, still growing. And most importantly, we deserve to know we're not alone in this journey. Meeting women traveling and living abroad brought me comfort and connection I had never anticipated, and now I aspire to offer that same support to others.

This isn't solely about travel or taxes; it's about creating a life that truly reflects who you are. Every woman deserves that opportunity.

Ultimately, I don't just assist women with filing taxes—I help them reclaim their time, peace, and power. When you understand how your money functions—especially as an American abroad—you stop living in fear, second-guessing your decisions, or feeling overwhelmed. Instead, you move forward with clarity and intention.

Take a deep breath. Embark on that journey. Ask the questions. Say yes to the life calling you. You're more prepared than you realize.

5.6 CONCLUSION—FROM MY HEART TO YOURS

Life is busy, and sometimes, we need to hear the full story before we can see ourselves in it.

If you're like me, you may have never envisioned living abroad, managing your own business, working remotely, or creating freedom on your own terms as something attainable. It wasn't part of the plan you were given.

I was taught to be responsible, work hard, take care of what I have, and be grateful. And I am. However, I also believed that gratitude didn't mean I had to stop striving for more.

Perhaps you feel this as well—the quiet sense of longing, the subtle whisper in the back of your mind when you're working at your laptop, walking into the same office day after day, or sitting at dinner listening politely while your mind wanders elsewhere.

You may be longing for:

- More freedom
- Greater options
- New adventures
- Increased control over your time, finances, and energy

If so, please know that you are not selfish, you are not unreasonable, and you are certainly not alone.

You may simply be ready—ready for a different path.

And here's a truth often left unspoken: even when you take the leap, purchase your ticket, pack your suitcase, and begin your new life abroad, there are still challenges many don't warn you about—especially around taxes.

Being an American abroad doesn't mean you've left all IRS regulations behind. In fact, the complexities often increase. The rules follow you across oceans, time zones, and currencies.

You're still required to file annually, regardless of where you are. You may have income in multiple currencies, foreign bank accounts to report, and special provisions such as the Foreign Earned Income Exclusion, the Foreign Tax Credit, or Foreign Bank Account Report (FBAR) requirements, which you might not even be aware of.

If you own a business, you might need to reassess whether your current structure remains suitable once you're living abroad.

Additionally, tax laws are always evolving, deadlines shift, and when you're in a different time zone, staying compliant can become even more complicated.

This can feel overwhelming and may diminish the joy of this new chapter if you don't have the right support.

I understand this firsthand. I've experienced the weight of trying to navigate these complexities alone. I know what it's like to worry about missing important details that could cost thousands. I understand the desire to enjoy your freedom but feeling buried under rules you didn't realize existed.

That's why I do what I do.

While I assist with practical matters—such as tax planning, compliance, business structuring, deductions, and navigating IRS regulations from abroad—that's only part of my mission.

My true goal is to help women feel secure, clear, and confident in the life they've chosen.

Because you deserve to be in control. You deserve peace of mind, knowing you've met all requirements and aren't stressing over missed deadlines or overlooked details. You should be able to focus on living your life, not just managing its logistics.

This isn't just theory for me. I am living this reality, managing foreign bank accounts, paying expenses in multiple currencies, swapping SIM cards, and deciding last-minute where I'll be next quarter.

When I say, "I get it," I mean it wholeheartedly.

This isn't merely work for me—it's personal. I understand what's possible when you do it right: the freedom to live boldly, travel extensively, and manage your finances without feeling tethered to one place or way of life.

You do not need to choose between freedom and stability. You can have both.

And I am here to support you every step of the way—helping you plan, pivot, and grow without the stress of taxes, confusion, or guilt about wanting more.

Because you deserve a life that feels authentic to you. And you don't have to navigate it alone.

When you're ready, I'll be here. Together, we can find your path forward.

5.7 CONCLUSION—YOUR LIFE, YOUR WAY

When I look back now, every twist, every delay, every "this is not what I planned" moment was leading me right here, helping women like you keep more of your money, feel confident about taxes, and live your life without second-guessing every step.

I've been that woman who didn't know taxes could be a tool for freedom.

The woman who opened a letter from the IRS and felt her stomach drop.

The woman who thought freedom and travel were for "other people."

But I've also been the woman who booked that first solo flight to Morocco and nearly talked herself out of going.

The woman who stood in an airport, heart racing, tears threatening, and almost turned back until a kind airline agent reminded me I was capable of more than I believed.

The woman who wandered the streets of a foreign city and realized, for the first time, she was exactly where she was supposed to be.

I've been the woman who learned to slow down in a café in Peru.

The woman who woke up one morning and realized she no longer measured her worth by how busy she was.

The woman who stopped chasing everyone else's definition of success and started building her own.

And now? Now I help other women do the same without the tax headaches, without the sleepless nights wondering if they messed something up, without having to choose between safety and adventure.

Here's what I know for sure:

You don't have to wait for the perfect time.
You don't need to have it all figured out.
You just have to want it enough to take the first step.

If you've ever felt that pull for more—more freedom, more peace, more control over your time and money, you're not crazy, you're not selfish, and you're definitely not alone.

Because I've been where you are. I've sat with the doubts. I've wondered if I was capable. I've worried about money, about security, about what people would say.

Because I've been where you are. I've sat with the doubts. I've wondered if I was capable. I've worried about money, about security, about what people would say.

But here's what I've learned: those fears don't disappear by staying where you are. They disappear when you take the first step and prove to yourself you can do hard things.

Your dreams aren't too big.
Your desires aren't unreasonable.
And wanting both stability and freedom doesn't make you unrealistic, it makes you human.
You deserve mornings that feel peaceful.

You deserve work that excites you.
You deserve adventures that make you feel alive.
You deserve to feel safe in your decisions and confident in your future.
So say yes to yourself.
Take the trip.
Start the business.
Make the move.

And when you do? I'll be here cheering you on, guiding you through the money and tax side, and reminding you every single step of the way that you've got this.

Because freedom isn't just about where you live. It's about how you live.
And you? You get to live it your way.
If I could leave you with just one truth, it's this:
The life you want isn't waiting for permission. It's waiting for you.
And the moment you decide to go after it? That's the moment everything changes.

6 Navigating Intersectional Challenges

How Ethnic Background Shapes Migrant Women's Entrepreneurial Journeys

Giulio Palmas

6.1 INTRODUCTION

While migrant entrepreneurship is pivotal for economic development and social and cultural integration in host countries (Van Riemsdijk & Basford, 2022; OECD, 2021), migrant women's entrepreneurship has emerged as a phenomenon of growing interest, generating economic opportunities, combating poverty, and facilitating social inclusion (Addo, 2017, De Luca & Ambrosini, 2019). Yet, despite substantial work on migrant entrepreneurship and on women's entrepreneurship, significant gaps remain in examining how ethnic and cultural backgrounds shape these experiences within a single national context (Halkias & Caracatsanis, 2016; Chreim et al., 2018).

In recent decades, enterprises initiated by migrants have significantly influenced host countries' economic and social development (Aman et al., 2021, 2022). However, despite growing interest in women-led ventures founded by migrants, comparative analyses across different ethnic and cultural groups within a single national setting remain scarce (Essers et al., 2010; Poggesi et al., 2016). Following Lassalle and Shaw (2021), understanding how diverse socio-economic and cultural backgrounds intersect with varying institutional structures is essential. For present purposes, we contextualize these dynamics in Italy and adopt an "intersectional" lens (Crenshaw, 1991) to grasp the interplay of gender, ethnicity, cultural background, and class.

Analytically, we examine how gender, ethnicity, cultural background, and class interact to shape the entrepreneurial trajectories of two Migrant Women Entrepreneurs (MWEs) in Italy. Methodologically, we draw on in-depth, semi-structured interviews within a case study design to assess obstacles to accessing financial resources, building business networks, and addressing cultural and institutional barriers, and we show how skills acquired in countries of origin combine with those developed in Italy to inform inclusive business models. This design enables a clear comparative account of individual experiences, shedding light on the intersectional dynamics that influence their entrepreneurial trajectories.

DOI: 10.1201/9781003604402-6

We address three research questions: (1) In what ways do gender, ethnicity, cultural background, and class intersect to shape the entrepreneurial trajectories of MWEs in Italy? (2) What challenges do MWEs face in accessing financial and social resources, and how do they address cultural and institutional barriers? (3) How do skills acquired in countries of origin combine with those developed in Italy to foster inclusive business models?

Our contribution is both theoretical and practical. Theoretically, we offer an empirically grounded account that advances debates on how gender, ethnicity, cultural background, and class intersect in shaping MWEs' entrepreneurial trajectories. Practically, we set out implications for the design of more effective policies and support programs tailored to the needs of this heterogeneous group.

The chapter is organized as follows. First, we set out the theoretical positioning and methodological approach, focusing on the "intersectional" lens and the qualitative, interview-based design. Subsequently, we present the case studies and data analysis, examining the different entrepreneurial trajectories. We then discuss the findings in relation to integration theory, practice, and policy. Finally, we draw conclusions and set out directions for future research, considering implications for both scholarship and practice.

6.2 THEORETICAL BACKGROUND

While women's entrepreneurship has emerged as a critical driver of innovation, job creation, and community development worldwide (Brush, 1992; De Bruin et al., 2007; Jennings & Brush, 2013), the share of women starting or running businesses remains lower than that of men (Elam et al., 2019), thereby reinforcing stereotypes such as "think entrepreneur–think man" (Laguía et al., 2022). From a problematization perspective (Alvesson & Sandberg, 2011), we suggest these perceptions originate in social constructions tied to masculine paradigms (Ahl, 2006). Moreover, contextual factors and individual capabilities tend to significantly frame entrepreneurial opportunities (Shane, 2003; Welter, 2011; Allen et al., 2007; Emami et al., 2022). Yet much research continues to treat women as a homogeneous group, overlooking distinctions related to ethnicity, migration status, and socio-economic background (Ahl & Marlow, 2012; Jennings & Brush, 2013).

Migrant women's entrepreneurship warrants particular analytical attention. Research suggests that these entrepreneurs may contribute to innovation and economic growth in host countries, while potentially enriching business ecosystems through cultural insights and transnational networks (Chreim et al., 2018; GEM, 2021). Such ventures tend to foster "diversity, equity and inclusion," elements increasingly regarded as significant for economic and social sustainability (Sternberg et al., 2023); furthermore, their success can generate what has been characterized as a "multiplier effect," potentially inspiring other women and strengthening social fabric (Shah & Saurabh, 2015). However, institutional barriers persist: notably, the non-recognition of homeland qualifications tends to relegate migrant women to underutilized roles, thereby constraining their entrepreneurial opportunity structures (Riaño, 2021).

Evidence suggests that these women encounter specific barriers: difficulties in accessing start-up finance and establishing both formal business networks and informal ties (Azmat, 2013; Sarihasan et al., 2023); "negative self-perceptions" and heightened "risk aversion." These constraints, compounded by structural barriers including sexism and racism (Goyal & Parkash, 2011; Chiang et al., 2013; Terjesen & Lloyd, 2015; Ram et al., 2017), further restrict their opportunities. At the individual level, deeply internalized home-country gender norms encounter more progressive host-country expectations; this tends to influence self-esteem and the development of entrepreneurial capabilities (Azmat, 2013; Maharjan et al., 2024). Yet adaptation is evident: many combine cultural resources from the country of origin with newly acquired capabilities in the Italian context to develop viable, and at times innovative, business models (De Luca & Ambrosini, 2019; Selcuk & Suwala, 2020). Alongside this, linguistic challenges and social isolation continue to impede access to information, formal business networks, and business support/advisory services (Lassalle & Shaw, 2021; Aman et al., 2024).

These network dynamics extend to other dimensions of migrant entrepreneurship. Family dynamics tend to be pivotal: the overlap between personal and professional spheres often imposes dual pressures; home-country norms shape entrepreneurial choices, while host-country expectations demand adaptation to competitive environments (Basco et al., 2019; Selcuk & Suwala, 2020). This can generate a "double burden," in which the pursuit of entrepreneurial success conflicts with the need to uphold traditional care and household responsibilities, thereby inhibiting entrepreneurial autonomy, especially when family support or alternative care arrangements are lacking.

Furthermore, the institutional and cultural context of the host country shapes migrant women's business environments. Barriers such as "xenophobia," language difficulties, and restricted access to formal business networks, alongside "patriarchal norms" and cultural expectations, tend to constrain their entrepreneurial trajectories (Solano, 2023; Aman et al., 2024). The lack of targeted entrepreneurship training, for instance specialized language courses, may compound these difficulties. A reflexive perspective (Alvesson & Sandberg, 2011) indicates that these conditions are not neutral; they reflect the host country's priorities and values. Studies suggest that in contexts with proactive support policies, the gender gap in entrepreneurship narrows, yet relying solely on individual initiative risks perpetuating structural barriers (Lassalle & Shaw, 2021; Kabbara et al., 2025). Furthermore, the success of such policies often depends on the active involvement of migrant women in their formulation (Alvesson & Sandberg, 2011; Brieger & Gielnik, 2021).

From this we deduce that an intersectional approach (Crenshaw, 1991) is vital to understanding how gender, ethnicity, cultural background, class, and migration status intersect to shape opportunities and constraints for migrant women entrepreneurs. In particular, the intersection of gender and ethnicity can foster the emergence of new entrepreneurial identities, offering channels for innovation and creativity (De Vita et al., 2014). This intersectional lens not only exposes systemic inequalities but also emphasizes the potential to harness transnational resources to inform targeted policies and support programs (Basco et al., 2019; Floris & Dettori, 2024).

To deepen understanding of migrant women's entrepreneurship in Italy, we adopt an intersectional perspective to examine how gender, ethnicity, cultural background, and class intersect to shape challenges related to (1) access to finance, (2) the formation of formal business networks and informal ties, and (3) the surmounting of institutional and cultural barriers. We also explore how capabilities acquired in home countries are combined with those developed in Italy, thus informing inclusive, context-specific support programs. Drawing on qualitative interviews with two migrant women entrepreneurs from different national backgrounds, the study aims to provide evidence to guide more targeted and effective policy interventions.

6.3 METHODOLOGY

This study uses a qualitative research design to explore the entrepreneurial experiences of migrant women in Italy. Given the emphasis on how gender, ethnicity, and cultural background intersect in shaping entrepreneurial trajectories, a semi-structured interview format was chosen (Guba & Lincoln, 1994). This method provides sufficient flexibility to explore unexpected topics as they arise, while maintaining a guiding structure to address the core research questions.

Participants were identified using a snowball sampling to ensure diversity in terms of country of origin, sector of activity, and years of entrepreneurial experience. The inclusion criteria required that (1) participants be first-generation migrant women, (2) they have established a business in Italy, (3) they possess at least three years of professional experience in their sector, and (4) they originate from different geographical regions. Although ten women in total were interviewed, this chapter focuses on two cases selected for their marked differences in cultural, professional, and personal backgrounds—thus illustrating the breadth of entrepreneurial paths among migrant women.

Estelle (51 years old) is a Franco-British entrepreneur who, after extensive international experience in teaching and educational management, settled in Sardinia around 2014. Bilingual from birth (English and French), she worked in various countries, including Kenya, Ukraine, and Cambodia, before relocating to Italy. Drawing on her language expertise, she established a small language school called "No Stress," specializing in flexible, communication-oriented English and French courses for adults and professionals. Estelle's story highlights the interaction of multicultural identity, language skills, and personal networks in shaping her entrepreneurial journey.

Gabriela (38 years old) is originally from São Paulo, Brazil, where she graduated with a Bachelor's degree in Sports Science. She moved to Italy in 2009 to join her then-husband, a professional football player, and initially faced language and contractual challenges in the sports and fitness sector. Her career took a decisive turn when she became a certified Zumba and fitness instructor, enabling her to move from precarious part-time roles to more stable, self-managed collaborations within a local sports federation. Today, Gabriela teaches several fitness classes, organizes public events, and is known for incorporating Brazilian cultural elements into her programs.

Data were gathered primarily through face-to-face semi-structured interviews, each lasting approximately 45–60 minutes. To encourage participants to provide detailed accounts, the interview guide began with questions about their personal trajectories—how they arrived in Italy and what motivated them to become entrepreneurs. This autobiographical prompt allowed them to reflect on both the triggers for their migration and the genesis of their entrepreneurial endeavors, creating a foundation for subsequent, more focused questions.

As the interviews proceeded, participants were asked about their education and professional experiences in their home countries, examining how their existing skill sets influenced—or sometimes complicated—their entrepreneurial activities in Italy. To explore financial aspects, they were invited to describe any difficulties encountered in securing funding or other resources, along with any biases or stereotypes they perceived when approaching banks or investors as migrant women.

Continuing this line of inquiry, the interviews then explored the formation of professional networks, including relationships with clients, suppliers, and collaborators. The interviewer encouraged participants to share specific examples of networking strategies—such as attending local markets or organizing culturally focused events—that helped them build valuable business relationships. At this point, both participants recalled community-based initiatives that simultaneously promoted social inclusion and professional opportunities.

In keeping with the intersectional lens of this study (Crenshaw, 1991), the interviews also explored how overlapping social identities—gender, migrant status, and socio-cultural background—shaped everyday business decisions, from pitching to investors to engaging with clients. Participants were asked whether any identities became more salient in particular scenarios and how they managed these overlapping roles in day-to-day operations.

In order to ensure a participant-driven narrative, the participants were encouraged to speak freely about all relevant experiences throughout the interviews. Open-ended prompts allowed for further exploration of the themes that emerged and provided a more complete picture of each woman's perspective. Interviews were recorded, fully transcribed, and carefully checked. The data were examined in iterative steps, following Braun and Clarke's (2006) approach to thematic analysis. The process began with familiarization—rereading and annotating the transcripts—before moving on to coding, which identified salient segments of text. These codes were then consolidated into broader themes reflecting recurring themes such as "access to finance," "network building," "intersectional challenges," and "balancing family and business." NVivo software facilitated transparent organization and systematic review of the data.

Secondary sources such as social media profiles, company websites, and local press coverage related to the participants' businesses were also analyzed by the researcher to increase reliability. This material was often in support of, or in addition to, the participants' accounts of marketing strategies, community involvement, and public perceptions. In the final phase, particular attention was paid to how the interplay of migrant status and gender shaped entrepreneurial outcomes, and the themes were refined to ensure internal coherence and conceptual clarity.

6.4 RESULT

6.4.1 Accessing Resources

The analysis of both cases showed that both women used personal resources and informal support to start their enterprises, with minimal formal finance. The patterns are in line with the findings of Poggesi et al. (2016), who highlight that women entrepreneurs, in particular migrants, are often unable to access institutional grants and financing due to a lack of awareness or bureaucratic complications.

- Estelle explains that she did not apply for any official grants or incentives, partly because she "*didn't even know there were such programmes for migrants.*" Instead, she covered her start-up costs (such as renting a combined home and office space) from her personal savings. She recalls: "*I didn't know if there were grants or anything for migrant entrepreneurs. I only used my own savings and found a house that can be doubled as a language school.*"
- Gabriela also describes a difficult period at the beginning of her career, saying that she was earning "*500 or 600 euros a month*" before discovering a more profitable niche in dance-based fitness classes. She recalls: "*With Zumba, I have gone up to 2,000 euros a year. It has changed everything, but it has also meant that I have had to invest in training and in marketing.*"

These self-financing practices not only demonstrate resourcefulness but also underscore a systemic gap in public support, as both participants acknowledged limited knowledge of potential public grants (Kabbara et al., 2025).

6.4.2 Building Professional Networks and Social Connections

In both cases, informal and community-based connections proved to be key catalysts for the establishment and expansion of their businesses. Such reliance on personal and professional networks resonates with the findings of Essers and Benschop (2007), who emphasize how migrant women often compensate for institutional gaps by drawing on close relationships and community-led initiatives.

- Estelle highlights how some of her first students, particularly one group of technicians, became an important support structure:

  > "I had these six engineers who were there for me for everything—to help me find a good accountant, hang the blackboard, moral support. From the start they believed in me." She also coordinated events on local exploration and language practice, explaining how that "people are much keener to learn English when they can experience something fun, like a guided tour or a tasting."

- For her part, Gabriela has used her cultural background to stand out in the fitness community. She recalls:

  > "By teaching dance and fitness in different gyms, I got to know everyone. Then I'd do events and people would remember me as 'the Brazilian'. The word spread quickly."

As time passed, she gained more and more credentials working with other trainers as well as the occasional work with the general Brazilian community in Sardinia and a devoted clientele.

These examples underline the critical role of trust-based networks not only for resource access but also for sustaining business growth.

6.4.3 Cultural and Institutional Barriers

While none of the respondents had unsolvable barriers, they each dealt with unique challenges stemming from their various backgrounds and English levels. These results are consistent with Goyal and Parkash (2011) and Ram et al. (2017), who observe that migrant entrepreneurs are often subject to a complex interplay of cultural bias and institutional barriers that complicate their success trajectories.

- Estelle notes that some neighbors in her block of flats seemed suspicious of her foreign status: "*Some neighbours didn't want to greet me; I saw them every day, but they didn't say anything. But in my work, being foreign was an advantage—they came to me because I'm a native speaker.*"

  She attributes her relatively smooth handling of bureaucratic tasks to her European background and fluency in English and French—illustrating how linguistic and educational capital can mitigate institutional barriers (Poggesi et al., 2016).
- Gabriela recalls moments of skepticism about her qualifications, especially for Pilates and other fitness disciplines that require specialized knowledge: "*One teacher told me I'd never be able to teach Pilates, that it wasn't for me. I cried so much, but it motivated me to prove her wrong.*"

While she initially struggled with the Italian language, she attended public language courses and eventually adapted her business model during the pandemic by offering outdoor classes, demonstrating an adaptive strategy to overcome cultural and institutional obstacles (Azmat, 2013).

6.4.4 Intersectional Dimensions: Gender, Ethnicity, and Socio-Cultural Background

Adopting an intersectional lens (Crenshaw, 1991) highlights how overlapping identities can act as both barriers and assets, as also suggested by Ahl and Marlow (2012), who argue that gender and ethnic identities intersect with class and migrant status to produce different entrepreneurial outcomes.

- Estelle uses her Franco-British background as a strategic asset in a region where demand for authentic English language teaching is high. She notes: "*People here are often curious about me: I'm French, I'm English, I'm not married, I run my own business. It's different from what they're used to.*"

  She does not perceive any major gender discrimination and attributes much of her success largely to clients "*looking for a native speaker.*"

- Gabriela faced stereotypes associated with her Brazilian heritage and initial lack of Italian language skills, but she turned these disadvantages into a distinct advantage by highlighting her cultural background in dance-based fitness classes. She explains: *"Being Brazilian was a big help in Zumba—people see it as exotic. But I also had to work extra hard to be taken seriously in more formal settings."*

While Estelle leveraged her linguistic capital, Gabriela's experience underscores the importance of adapting business models to overcome cultural challenges.

6.5 SUMMARY AND IMPLICATIONS

Overall, Estelle and Gabriela's experiences highlight the central role of personal initiative, informal networks, and cultural identity in shaping migrant women's entrepreneurship, echoing Ahl's (2006) critique of the predominantly masculine paradigms that have long guided entrepreneurial discourse. Neither participant accessed formal financial support, citing a lack of awareness or insufficient incentives—confirming the observations of Poggesi et al. (2016) and Terjesen and Lloyd (2015) regarding the underutilization of external funding among women entrepreneurs, particularly those from migrant backgrounds. Instead, both relied on community-based support and self-funded professional development.

Their trajectories also highlight how intersecting identities—encompassing gender, ethnicity, and socio-cultural background—act as both barriers and differentiators. This observation reaffirms the utility of intersectional frameworks (Crenshaw, 1991; Ahl & Marlow, 2012) for understanding the nuanced realities of migrant women entrepreneurs. Moreover, it suggests that tailored policy interventions and training programs—providing language support, sector-specific mentorship, and accessible financing channels—could unlock additional potential for women like Estelle and Gabriela, who rely largely on personal resourcefulness and social capital to overcome systemic gaps.

Future research should explore larger samples and comparative analyses across different national contexts to further validate these insights and inform more inclusive support programs.

6.6 CONCLUSION

This study set out to understand the entrepreneurial dynamics of migrant women in Italy by adopting an intersectional approach, examining the interaction between gender, ethnicity, cultural background, and social class. Based on qualitative interviews with two different case studies—Estelle, a Franco-British entrepreneur running a language school, and Gabriela, a Brazilian fitness instructor—it provides differentiated insights into how these women navigate both opportunities and structural barriers in their entrepreneurial journeys.

The findings show that both Estelle and Gabriela relied primarily on their own resources and informal support networks to develop and sustain their business initiatives, rarely turning to formal funding or institutional support. This self-financing strategy, which they supported through ongoing personal development work, highlights the important part that social capital plays in overcoming financial and bureaucratic barriers. In addition, both entrepreneurs made effective use of their unique cultural identities: Estelle used her bilingual and multicultural background to enhance the appeal of her language school, while Gabriela turned initial skepticism about her qualifications into a competitive advantage in the fitness market by integrating Brazilian cultural elements into her offerings.

The important role of professional networks and community ties has also been a focus of the study. Informal relationships, whether through former students or local initiatives in the community, proved essential in mitigating the challenges of xenophobia, language barriers, and limited access to formal credit. These findings are in line with previous research (e.g., Essers & Benschop, 2007; Poggesi et al., 2016) demonstrating that trust networks are important for women's entrepreneurship, especially among migrant populations.

Furthermore, the intersectional approach reveals that the overlapping identities of these women act as both barriers and differentiators. While structural and cultural barriers such as bureaucratic complexity and subtle social exclusion persist, these entrepreneurs integrate skills from their countries of origin with those acquired in Italy to develop inclusive business models that respond to local market demands.

The small sample size restricts the generalizability of the findings, although they provide important evidence. For this reason, future research should broaden the study sample by considering different national contexts in order to explore the multiple issues and opportunities faced by migrant women entrepreneurs.

Estelle's and Gabriela's stories illustrate how individual initiative, cultural identity, and informal networks are crucial determinants of entrepreneurial outcomes. These insights contribute to the existing literature by reaffirming the relevance of intersectional frameworks (Crenshaw, 1991; Ahl & Marlow, 2012) and suggest that targeted policy interventions, such as tailored language support, sector-specific mentorship and simplified access to finance, could further empower migrant women and enhance both their economic and social integration in host countries.

REFERENCES

Addo, P.-A. (2017). 'Is it entrepreneurship, or is it survival?': Gender, community, and innovation in Boston's black immigrant micro-enterprise spaces. *Societies*, 7(3), 20. https://doi.org/10.3390/soc7030020

Ahl, H. (2006). Why research on women entrepreneurs needs new directions. *Entrepreneurship Theory and Practice*, 30(5), 595–621.

Ahl, H., & Marlow, S. (2012). Exploring the dynamics of gender, feminism and entrepreneurship: Advancing debate to escape a dead end? *Organization*, 19(5), 543–562.

Allen, I. E., Langowitz, N., & Minniti, M. (2007). Global entrepreneurship monitor. 2006 report on women and entrepreneurship, 3(1), 54–88.

Alvesson, M., & Sandberg, J. (2011). Generating research questions through problematization. *Academy of Management Review*, 36(2), 247–271.

Aman, R., Ahokangas, P., & Zhang, X. (2021). Migrant women entrepreneurs and entrepreneurial ecosystems during an external shock: A case study from the healthcare sector in Kazakhstan. *Asian Business & Management*, 20, 518–548.

Aman, R., Ahokangas, P., Elo, M., & Zhang, X. (2022). Migrant women entrepreneurship and the entrepreneurial ecosystem. In *Disadvantaged Entrepreneurship and the Entrepreneurial Ecosystem* (pp. 87–119). Emerald Publishing Limited.

Aman, R., Elo, M., Ahokangas, P., & Zhang, X. (2024). Empowering migrant women's entrepreneurship: Stakeholder perspectives from the entrepreneurial ecosystem. *International Journal of Entrepreneurial Behavior and Research* 30(7), 1698–1724.

Azmat, F. (2013). Opportunities or obstacles? Understanding the challenges faced by migrant women entrepreneurs. *International Journal of Gender and Entrepreneurship*, 5(2), 198–215.

Basco, R., Calabrò, A., & Campopiano, G. (2019). Transgenerational entrepreneurship around the world: Implications for family business research and practice. *Journal of Family Business Strategy*, 10(4), 100249.

Braun, V., & Clarke, V. (2006). Using thematic analysis in psychology. *Qualitative Research in Psychology*, 3(2), 77–101.

Brieger, S. A., & Gielnik, M. M. (2021). Understanding the gender gap in immigrant entrepreneurship: A multi-country study of immigrants' embeddedness in economic, social, and institutional contexts. *Small Business Economics*, 56(3), 1007–1031.

Brush, C. G. (1992). Research on women business owners: Past trends, a new perspective and future directions. *Entrepreneurship Theory and Practice*, 16(4), 5–30.

Chiang, F., Low, A., & Collins, J. (2013). Two sets of business cards: Responses of Chinese immigrant women entrepreneurs in Canada and Australia to sexism and racism. *Cosmopolitan Civil Societies: An Interdisciplinary Journal*, 5(2), 63–83.

Chreim, S., Spence, M., Crick, D., & Liao, X. (2018). Review of female immigrant entrepreneurship research: Past findings, gaps and ways forward. *European Management Journal*, 36(2), 210–222.

Crenshaw, K. (1991). Mapping the margins: Intersectionality, identity politics, and violence against women of color. *Stanford Law Review*, 43(6), 1241–1299.

De Bruin, A., Brush, C. G., & Welter, F. (2007). Advancing a framework for coherent research on women's entrepreneurship. *Entrepreneurship Theory and Practice*, 31(3), 323–339.

De Luca, D., & Ambrosini, M. (2019). Female immigrant entrepreneurs: More than a family strategy. *International Migration*, 57(5), 201–215.

De Vita, L., Mari, M., & Poggesi, S. (2014). Women entrepreneurs in and from developing countries: Evidences from the literature. *European Management Journal*, 32(3), 451–460.

Elam, A. B., Brush, C. G., Greene, P. G., Baumer, B., Dean, M., Heavlow, R., & Association, G. E. R. (2019). Women's entrepreneurship report 2018/2019.

Emami, A., Ashourizadeh, S., Sheikhi, S., & Rexhepi, G. (2022). Entrepreneurial propensity for market analysis in the time of COVID-19: Benefits from individual entrepreneurial orientation and opportunity confidence. *Review of Managerial Science*, 16(8), 2413–2439.

Essers, C., & Benschop, Y. (2007). Enterprising identities: Female entrepreneurs of Moroccan or Turkish origin in the Netherlands. *Organization Studies*, 28(1), 49–69.

Essers, C., Benschop, Y., & Doorewaard, H. (2010). Female ethnicity: Understanding Muslim immigrant businesswomen in the Netherlands. *Gender, Work and Organization*, 17(3), 320–339.

Floris, M., & Dettori, A. (2024). "With our feet on the ground and our minds free to fly": Multiple embeddedness and entrepreneurial orientation in small and medium-sized family businesses. *Journal of Management and Governance*, 28(2), 565–595.

GEM (2021). Global entrepreneurship monitor. *Global Enterpreneuship Monitor*, 51(8), 51-4537.

Goyal, M., & Parkash, J. (2011). Women Entrepreneurship in India: Problems and Prospects. *International Journal of Multidisciplinary Research*, 1(5), 195–207.

Guba, E. G., & Lincoln, Y. S. (1994). Competing paradigms in qualitative research. In N. K. Denzin & Y. S. Lincoln (Eds.), *Handbook of Qualitative Research* (pp. 105–117). Sage Publications.

Halkias, D., & Caracatsanis, S. M. (2016). The evolution of researching female immigrant entrepreneurship: A commentary. In Daphne Halkias, Sylva M. Caracatsanis (eds.) *Female Immigrant Entrepreneurs* (pp. 3–8). Taylor and Francis group.

Jennings, J. E. & Brush, C. G. (2013). Research on women entrepreneurs: Challenges to (and from) the broader entrepreneurship literature. *The Academy of Management Annals*, 7(1), 663–715.

Kabbara, D., Suárez-Ortega, S. M., & Zucchella, A. (2025). Developing entrepreneurial opportunities through multi-layered liabilities: The experiences of female migrant entrepreneurs. *International Entrepreneurship and Management Journal*, 21(1), 1–32.

Laguía, A., Wach, D., Garcia-Ael, C., & Moriano, J. A. (2022). Think entrepreneur–think male: The effect of reduced gender stereotype threat on women's entrepreneurial intention and opportunity motivation. *International Journal of Entrepreneurial Behavior & Research*, 28(4), 1001–1025.

Lassalle, P., & Shaw, E. (2021). Trailing wives and constrained agency among women migrant entrepreneurs: An intersectional perspective. *Entrepreneurship Theory and Practice*, 45(6), 1496–1521.

Maharjan, A., Nazir, M. A., & Roomi, M. A. (2024). Exploring the intersectionality of ethnicity, gender and entrepreneurship: A case study of Nepali women in the United Kingdom. *International Journal of Gender and Entrepreneurship* https://doi.org/10.1108/IJGE-01-2024-0009

OECD/European Commission. (2021). *The Missing Entrepreneurs 2021: Policies for Inclusive Entrepreneurship and Self Employment*. OECD Publishing. https://doi.org/10.1787/71b7a9bb-en

Poggesi, S., Mari, M., & De Vita, L. (2016). What's new in female entrepreneurship research? Answers from the literature. *International Entrepreneurship and Management Journal*, 12, 735–764.

Ram, M., Jones, T., & Villares-Varela, M. (2017). Migrant entrepreneurship: Reflections on research and practice. *International Small Business Journal*, 35(1), 3–18.

Riaño, Y. (2021). Highly skilled migrant and non-women migrants and men: How do differences in quality of employment arise? *Administrative Sciences*, 11(5), https://doi.org/10.3390/admsci11010005

van Riemsdijk, M., & Basford, S. (2022). Integration of highly skilled migrants in the workplace: A multi-level framework. *Journal of International Migration and Integration*, 23(2), 633–654.

Sarihasan, I., Dajnoki, K., & Al-Dalahmeh, M. (2023). Immigrant entrepreneurship and gender dimensions: A systematic review. *Intangible Capital*, 19(2), 110–130.

Selcuk, G., & Suwala, L. (2020). Migrant family entrepreneurship: Mixed and multiple embeddedness of transgenerational Turkish family entrepreneurs in Berlin. *Journal of Family Business Management*, https://doi.org/10.1108/JFBM-03-2019-0011

Shah, H., & Saurabh, P. (2015). Women entrepreneurs in developing nations: Growth and replication strategies and their impact on poverty alleviation. *Technology Innovation Management Review*, 5(8), 34.

Shane, S. A. (2003). *A General Theory of Entrepreneurship: The Individual-Opportunity Nexus*. Edward Elgar Publishing.

Solano, G. (2023). A level playing field for migrant entrepreneurs? The legal and policy landscape across EU and OECD countries. *International Migration*, 61(2), 27–47.

Sternberg, R., et al. (2023). Global Entrepreneurship Monitor: Unternehmensgründungen im weltweiten Vergleich, L€anderbericht Deutschland, 2022/23 GEM/RKW Kompetenzzentrum.

Terjesen, S., & Lloyd, A.. (2015). *The 2015 Female Entrepreneurship Index: An Analysis of the Conditions That Foster High-potential Female Entrepreneurship*. Washington, DC: Global Entrepreneurship and Development Institute.

Welter, F. (2011). Contextualizing entrepreneurship—Conceptual challenges and ways forward. *Entrepreneurship Theory and Practice*, 35(1), 165–184.

7 Women Leaders Driving Innovation in America

Case Studies on Overcoming Barriers and Shaping the Future

Harleen Kaur, Rosy Singh, Tilottama Singh and Melissa Hyunh

7.1 INTRODUCTION

Innovation in the USA is progressively being driven by women in leadership roles. Women leaders are making an increasingly important impact as founders of start-ups or creators of public policy, with different approaches to value creation, sustainability, and equity in mind. This study not only includes a few US-born women, but also a rapidly growing group of women expatriates who are foreign-born, who have migrated to the USA, and who are now leading innovation across technology, healthcare, education, and green industries. These women expatriates are often involved in navigating two systems constantly—cultural adaptation and discrimination—while they continue to emerge as disruptors and institutional change agents. Nevertheless, even with this increasing involvement, women remain disadvantaged in having access to venture capital funding, executive leadership roles, and innovation stories (Brush et al., 2020; Kamberidou, 2020).

Research highlights that the startups founded by women are granted disproportionately unequal share of funding from venture capitalists owing to continuing biases of investors concerning a founder's ability to handle risk and their past accomplishments, among other things (Kanze et al., 2018; Ewens and Townsend, 2020). However, the empirical evidence shows that women-run firms outperform their counterparts along multiple dimensions and have the potential for exceptional innovation (Marques & Ferreira, 2021). Women's entrepreneurial confidence is changing, with more women interpreting opportunities for entrepreneurship at the same time acknowledging fears of failure (Bosma et al., 2023). Women from varied ethnicities and immigrant entrepreneurs are contributing increasingly to the diversity of US entrepreneurial landscape. However, systemic barriers persist, as many minority entrepreneurs are reluctant to seek funding in advance out of a presaged sense of bias, even when creditworthiness and local creditworthiness are possible and present (Fairlie & Robb, 2007).

 DOI: 10.1201/9781003604402-7

More recently reported indicators about women entrepreneurs reflect some of this change. During 2022, 47% of new entrepreneurs in the USA were women, up from 29% pre-pandemic (World Economic Forum, 2023). Globally women represent nearly a quarter of driven entrepreneurs worldwide, with even greater representation in low-income nations (33%) and in North America at nearly 40% (Bosma et al., 2023). Even with those gains, women still lack representation in capital-intensive industries, where availability of funding and access to networks of influence are critically necessary. This all shows in the Pitchbook (2023) data, revealing that female-only founding teams obtained only 2.1% of total VC funding. At the same time, companies with 30% of women in leadership see up to 25% improvements in profitability, affirming the overall strategic value of diversity and inclusive leadership in impactful sectors, such as ICT, clean energy, and industrial tech (Brush et al. 2020; McKinsey & Company, 2020; NSF, 2023).

Despite this growth, traditional gendered expectations and persistent institutional exclusion hampers women's ability to fully participate. Women innovators' (principally expat and minority women) experiences often impose intersecting barriers to participation that include social expectations, inequities, and underrepresentation in patent ownership and S&E occupations (Houser et al., 2021; NSF, 2023). When women do lead, they are asked more questions about risk than discussing potential, and this framing raises questions about credibility and access (Kanze et al., 2018). Yet women continue to recreate the paths to innovation, through purpose-driven and community-centric leadership. Also the evidence suggests that women founders are more inclined to establish socially responsible businesses that align with ESG values and socially inclusive economic goals (Terjesen et al., 2015).

This research explores the structural barriers and enablers defining the trajectories of women innovators in America, explicitly on expatriate women leaders. We draw on a case study of several trailblazers to develop ideas regarding how women leverage networks, identity, and resilience to build transformational enterprises. Based on this evidence, the research suggests ecosystem reforms, including funding equity, gender-responsive funding, inclusive laws and regulations, mentorship pipelines, and governance policies grounded in equity. In the end, we hope to supe up a template for institutional change that doesn't just accommodate women but empowers them—regardless of place of origin—to lead the future of innovation.

7.2 HISTORICAL CHALLENGES FACED BY WOMEN

Women's participation and influence in American politics and economic sphere is a trajectory rooted in a long history of systemic subordination. American women transitioned from having no rights to vote, to gaining those rights, particularly through the ratification of the 19th Amendment in 1920, after decades of sustained suffrage activism (Flexner & Fitzpatrick, 1996). For the majority of the 19th and 20th centuries, hegemonic patriarchal norms meant that women were largely forced to occupy domestic roles, stripping them of equitable opportunities for education, professional opportunity, and thus voice in civic society. This structural bias is most notable in the fields of "science, technology, engineering and mathematics (STEM)," where women, as a consequence of being denied equitable opportunity, participated in

STEM occupations at 8% of the workforce in America in 1970 (U.S. Census Bureau, 2021). Similar to STEM, and with respect to positions of leadership in politics, medicine, and business (corporate management, entrepreneurship), social norms disadvantaged women to the extent where women were deprived voting rights, property ownership, and even livelihood as a means to sustain financial independence (Purkayastha, 2020) (Table 7.1).

Despite systemic gendered inequities, women catalyzed political and social movements throughout the 20th century that resulted in transformative political and social change, which collectively resulted in key legislative initiatives that represented significant turning points because they represented the institutionalization of women's inclusion. Among many important legislative initiatives was the 1963 parity in wages for women workforce. The Equal Pay Act, the first federal law to prohibit pay discrimination by sex, resulted in the legislative floor for the equal pay debate. Despite the prohibition of pay discrimination by gender, many working women benefited from this legislation to strengthen the prevalence of women working outside of the domestic realm (U.S. Department of Labor, 2023). Another major legislative act was Title IX Education Amendment Act of 1972, which ensured equal access to educational programs, activities, benefits, and employment opportunities, regardless of sex. Title IX had significantly expanded the opportunities for women in athletic and academic programs, fostering an increase in women's role as leader and innovators across all sectors (Buechler, 1990).

As the 21st century began, women secured a formidable position as catalysts of social and cultural change in all spheres of life and fundamentally distanced themselves from hegemonic gender normativity. In 2023, women accounted for approximately 41% of new entrepreneurs in the USA, demonstrating their increasing presence in entrepreneurship and business leadership (Kauffman Foundation, 2023). Even though women are underserved in some STEM disciplines, they earned the most degrees in some areas of study, including biological sciences, especially among those under 30 years old (NSF, 2023). These trends demonstrate how there has been

TABLE 7.1
Key US Policies Advancing Women's Rights and Participation in STEM and Innovation

Year	Policy/Act	Area of Impact	Outcome for Women
1920	19th Amendment	Political Rights	Women gained right to vote
1963	Equal Pay Act	Workplace Equality	Legal foundation for pay parity
1972	Title IX	Education/STEM	Boosted female STEM participation
2022	CHIPS and Science Act	Tech & Innovation	Inclusion of women in STEM R&D funding

Source: Compiled from historical legislation and public records, including the U.S. National Archives, U.S. Department of Labor, U.S. Department of Education, and the CHIPS and Science Act (2022) as documented in the U.S. Congress Legislative Information System.

a generational mindset shift that has opened up opportunities for women—in both educational and professional sectors—throughout the country. This chapter presents a collection of case studies of women leaders and innovators who have not only successfully navigated systemic barriers to access and equity, but who have further redefined hegemonic norms of leadership, entrepreneurship, and innovation within an American context.

7.3 CASE OF WOMEN INNOVATORS IN BUSINESSES

In the evolving corporate landscape in the USA, a new cadre of feminine leadership has emerged who are not only transforming organizational strategy but also promoting sustainability, technological change, and inclusive leadership. This section will highlight the women who are leaders of innovation-driven transformation in traditionally male-dominated industries.

7.3.1 Indra Nooyi—Former CEO and Chairperson, PepsiCo (2006–2018)

Indra Nooyi has received global recognition as an Indian-American business executive for redefining corporate leadership with a purpose-driven approach. Nooyi broke multiple glass ceilings, becoming the first Indian-origin female chief executive of a Fortune 500 company, and leading a model of executive leadership that incorporated strategic growth with responsible business management (Magnanelli & Raoli, 2025).

7.3.1.1 Leading at PepsiCo

After joining PepsiCo in 1994, Nooyi assumed the role of Chief Executive Officer (CEO) in 2006. Her most famous initiative was called Performance with Purpose. This initiative emphasized the idea of sustainable product innovations, healthy food items, and engagement with stakeholders (Santos et al., 2021). She ensured that PepsiCo was aligned with global trends through:

- Repositioning the product portfolio to be more nutritional and wellness-driven.
- Expanding into emerging markets, to ensure global sustainability.
- Encouraging ESG values in the culture of the company.

Nooyi generated both financial capital and reputational capital in her leadership position, showing all-inclusive capitalism can lead to success in a competitive market (Kiron et al., 2017).

7.3.1.2 Systemic Challenges

Nooyi was a woman of color at an FMCG brand, which placed her into systemic barriers and to experience what researchers have identified as a "glass cliff" phenomenon, where women are often encouraged to take on leadership roles amid the periods of decline or turbulence (Ryan & Haslam, 2005). To elaborate further, the consumer goods industry, particularly in North America, has been slow to advance women into leadership roles at the firm level in the top tier (McKinsey, 2021). Nooyi had to also consider family expectations and cultural dimensions as a woman occupying a male-dominated executive space.

7.3.1.3 Strategic Innovation and Potential Impact

Nooyi introduced the Performance with Purpose platform, a transformational journey to the corporation, that meant to embrace sustainable principles, health-conscious product lines, and employee-first policies. Nooyi steered the growth in revenues of PepsiCo, which in 2006 were $35 billion and later in 2017 rose to $63.5 billion (PepsiCo Annual Report, 2017). Nooyi also was pioneering an investment into sustainable packaging, a commitment to reduce PepsiCo's carbon footprint, and lined up growth into a healthier portfolio of food and beverage alternatives, in line with the global health trends (Nooyi, 2021). Nooyi's leadership revolutionized the concept of CSR (corporate social responsibility), and positioned PepsiCo as a leader in sustainability.

7.3.2 Dr. Lisa T. Su—President and CEO, Advanced Micro Devices (AMD)

Dr. Lisa Su immigrated from Taiwan to New York and went on to become the first Taiwanese woman to helm a publicly traded US semiconductor business. Dr. Su began her career with positions at Texas Instruments. During the early stages of her career, Dr. Su assumed the role of Vice President at IBM semiconductor research and development center and subsequently was elevated to General Manager at IBM semiconductor product center. Advanced Micro Devices (AMD) hired Dr. Su in January 2012 where she went on to become the Senior Vice President.

Dr. Lisa Su began her tenure as AMD's CEO in October 2014 and has since led AMD through a monumental transformation from an at-risk company to a technology titan. When Dr. Su became CEO of AMD, its stock was only 2.79/share, while the price was 174.43/share in October 2024, a staggering increase of 6,151.97%. This means that an investment of $1,000 at the beginning of Dr. Su's tenure would now equal $62,519.71. AMD was generally viewed as a turnaround underdog with plenty of problems, but with upside potential for growth. Meaningful milestones during Dr. Su's tenure include the acquisition of Xilinx for $35 billion in 2022, which doubled EPYC processors' sales and expanded the company's total addressable market (TAM) from $80 billion to $135 billion. In addition, AMD completed a share buyback totaling $10 billion and increased shareholder value by 6.4%, as described by AMD stock price growth, following its completion in May 2025. This disciplined capital allocation strategy, accompanied with innovation expansion and aggressive acquisitions, not only transformed AMD but also created an industry standard for growth, operational excellence, and shareholder returns.

Dr. Su's success at AMD has brought the company the respectable ranking of 181 on the most recent Fortune 500 list, making her one of among 55 women of color as CEO of America. Dr. Dr. Su's successful narrative carries the influence of an immigrant, particularly in the context of Taiwan and current dominant role in the global semiconductor supply chain. Dr Su has successfully leveraged this position to build domain partnerships in the chip space from Asia and the USA. Dr. Su's skill set of technical virtue, cultural intelligence, perceptiveness, and data-driven talent insights constantly drives her against her intentions to enhance and overcome the unrelenting challenges facing her under AMD. Perhaps this is why expatriate women leaders are capable of this delivery of extreme innovation, renewal of lost market share, and shareholder value for companies in America (Kumari & Nair 2023).

7.3.3 Ursula Burns—Changing Corporate America

Ursula Burns made history in 2009 when as a minority woman she rose to the ranks of CEO of Xerox. Born in 1958, Burns grew up in public housing on the lower east side of Manhattan, being raised by her single mother (Rholetter, 2024). Equipped with a Bachelor's in mechanical engineering from NYU and a Master's from Columbia University, Burns joined Xerox in 1980 for a minority engineering internship. From there, she continued to advance the company in engineering and product development (Burns, 2021).

7.3.3.1 Challenges Faced

Burns's advancement negated two major corporate biases, gender and race. The research confirms the theory that Black women are often described as facing "intersectional invisibility," which suggests they are marginalized in organizations more than others of their gender and of their ethnicity (Purdie-Vaughns & Eibach, 2008). Burns's appointment came during a time when Xerox was viewed as a legacy business on the fringe of digital transformation. Many analysts feared that Xerox would lapse into irrelevance as demand for traditional copiers and printers plateaus or declines into nothingness (Knowledge@Wharton, 2010).

7.3.3.2 Innovation and Impact

Burns drove transformation at Xerox by securing a $6.4 billion acquisition of Affiliated Computer Services (ACS) in 2010, moving the company away from total reliance on hardware. By 2012, revenues from services had doubled and accounted for more than 50% of total revenues (Forbes, 2012). Burns enacted "transformational innovation" and altered the company's prospects for survival in three ways: firstly, team diversity raised the pool of ideas and challenges; second, flexible adaptation to improve outcomes; and third, thinking into the future as strategic human choice (Hossain, 2016). Burns' success opened symbolic capital for future leaders of the future, but it also redefined accountability for global boardroom diversity.

7.4 EXPAT WOMEN INNOVATORS IN TECHNOLOGY GLOBAL CONTEXT

7.4.1 Women Expatriates Reshaping the Tech Frontier in 2025

While strides have been made, the gender disparity within the global tech ecosystem is still immense. Women represented 25–30% of the total technology workforce globally in 2025, in addition to holding less than 20% of senior leadership positions

(World Economic Forum, 2024; McKinsey Global Institute, 2025). Furthermore, this decreased when we consider data available on expatriate assignments—women held only 14% of international assignments in that industry (PwC, 2024). While India's IT sector has reported gender parity gains in senior leadership roles—23% senior female leadership in 2024 from 18.7% in 2023 (NASSCOM, 2024)—women globally in the technology sector with expatriate placements continue to confront systemic barriers in cultural bias, sponsorship, and/or being assigned to critical, high-visibility projects. There remain chronic gender pay inequities and challenges related to retention and promotion, which have resulted in expectations of gender parity in representation and pay in many economies to still be 20–30 years away, unless deliberate systemic reforms speed up timelines (OECD, 2024).

7.4.2 Expat Women Pioneering Data-Driven Innovation

One of the best-known women expat's pioneers and innovators on the frontiers of artificial intelligence (AI) is Mira Murati, an Albanian-born, mechanical engineer with an MIT education who was a key contributor to advanced projects at Tesla and Leap Motion before pivoting to AI. From 2018 to 2024, as Chief Technology Officer at OpenAI, Murati directed the global development and launch of generative AI products—ChatGPT, DALL-E, and GPT-4—to global-scale user adoption before ChatGPT was used by over 100 million users within 2 months after launch. During an organizational restructuring at OpenAI, she quickly became interim CEO for 2023, garnering praise for her strategic clarity and navigation through crisis. Murati publicly acknowledged gender gaps in AI, including the 22% female participation in technical roles within OpenAI, and she aligned her leadership agenda toward developing new talent pipelines of diverse participants. In 2025, she launched a new venture Thinking Machines Lab, to democratically bring access to interpretable AI systems, and globally build diverse cohorts of researchers to particularly draw from expatriate female technologists underrepresented in technology (National CIO Review, 2025).

Another influential figure is Dr. Fei-Fei Li, who relocated to the USA in 1995 from China and co-founded the Stanford Human-Centered AI Institute. Dr. Li is most recognized for her work on ImageNet, a project that underscored the proliferation of deep learning by training large swathes of image recognition models, which have been cited in 100,000s of scholarly papers. While serving as Chief Scientist of AI at Google Cloud, she focused on the responsibility of developing AI frameworks while building equitable algorithmic governance across both public and private sectors. In 2015, she founded AI4ALL, a non-profit that focuses on advancing AI education to historically marginalized youth (Li, 2024). By 2024, AI4ALL had worked with over 4,000 high school students across 20+ states in the USA and internationally focused on expat, immigrant, and refugee girls' pathway into STEM. Her advocacy at the United Nations and the United States Congress taxing the exact definition of equitable action reflects her desire for policy quotient, ultimately when she was recognized as one of Fortune's World's 50 Most Powerful Women Leaders.

7.5 CASE OF EXPAT WOMEN INNOVATORS IN AMERICAN HEALTHCARE

The US healthcare system has been bolstered by many immigrant women who offer cross-border expertise, multilingual abilities, and a transnational mindset to local medical problems. In other words, immigrant women are often working in the healthcare system as frontline workers, scientific researchers, and policy developers. All of these roles help to support the important goals of clinical efficiency and equity in public health. These women improve health outcomes, progress vaccine innovation, and change medical education, but their journeys also reveal systemic inequities related to immigration status, wage inequities, and the lack of representative voices in leadership. This section highlights some of the important leaders whose innovations in science and healthcare continue to shape and develop American healthcare, especially during the unprecedented time of the COVID-19 pandemic.

7.5.1 Expat Women Transforming US Healthcare

The trajectory of modern US healthcare is interwoven with the stories of women whose roots lie as US nationals and expatriates. From the profile of workers in the healthcare system, it was noted that in 2021, approximately 75% of the estimated 2.8 million immigrant workers were women, spanning occupations from direct care to advanced biomedical research (Migration Policy Institute, 2022). This group includes many physicians, nurses, public health officials, and clinical researchers. Importantly, they are often working in underserved communities, where the multitude of languages, cross-cultural understanding, and global medical training ensure that clinical outcomes and patient satisfaction metrics are raised (NIH, 2023). These women provide critical frontline care—during the COVID-19 pandemic—and serve as bridges between multicultural patient populations and western medicine, aiding in improving health equity in structurally marginalized communities.

Concurrently, immigrant women in healthcare are impacted by complex and layered limitations: limitations from visa issues, lack of access to tracks for leadership positions in health organizations, underrepresentation in medical boards, and minimum wages. For example, immigrant women nurses earn 7–14% what native-born nurses make when the immigrant women nurses' clinical performance is rated the same or higher, even sometimes cited for praise (American Nurses Association, 2023). In particular, the green card backlog occurs when skilled female health workers from Africa and Asia are delayed in their career development and their research (USCIS, 2024).

7.5.2 Dr. Özlem Türeci—Physician-Scientist and Biotech Pioneer

Dr. Özlem Türeci moved to Germany from Turkey and changed the landscape of US biomedical innovation as Co-Founder and Chief Medical Officer of BioNTech, the European firm that enabled the creation of the mRNA vaccine for COVID-19. Both a practicing oncologist and immunologist, Dr. Türeci's unique training and experience positioned her as a critical link to guide clinical knowledge to bioengineering precision. Between 2020 and 2022, she also co-led several research initiatives catalyzing vaccine deployment at an unprecedented global scale, including within the USA, where the Pfizer-BioNTech mRNA vaccine became part of the public health strategy: the vaccine was administered to more than 350 million Americans over the first 18 months after mRNA vaccine approval (CDC, 2022).

Dr. Türeci's work exemplifies translational medicine, where research concepts in the laboratory must be pivoted quickly for public health implementation, and she continues to be a voice for vaccine equity for under-resourced and marginalized communities. Dr. Türeci has also been a strong advocate for women in clinical research, with BioNTech reporting a 41% increase in women researchers from 2020 to 2023 (BioNTech, 2024). Dr. Türeci has received several accolades regarding her role in the pandemic, including TIME naming her one of the "100 Most Influential People" for 2021. Her work explains how expatriated women fill the gap between national health security and global scientific diplomacy (Deutschland.de, 2021).

7.5.3 Dr. Joanne Liu—Emergency Health Leader and US Policy Innovator

Dr. Joanne Liu, a Chinese-Canadian physician educated at McGill University and Harvard School of Public Health, has had a significant influence on US emergency response protocols through her global humanitarian authorities. She was International President of Médecins Sans Frontières (MSF) from 2013 to 2019, guiding humanitarian missions globally, including leading and advising on the West African Ebola outbreak that directly impacted pandemic response frameworks utilized in American hospitals during the COVID-19 pandemic response, specifically in New York and Boston (McGill University, 2021).

As a faculty member of Harvard Medical School and NYU Grossman School of Medicine, Dr. Liu has led and developed global emergency health curricula, impacting over 2,000 students and fellows annually (WomenLift Health, n.d.). She could utilize her multiculturalism and field experience to focus on multinational case triage systems and approaches to US emergency department settings, through community-based outbreak response while developing culturally informed care and language access using multilingualism. Her US Senate testimony on COVID-19 in 2021 was utilized with both US Republican and US Democrat legislators in drafting bipartisan legislation on pandemic preparedness, including the Global Health Security Act (Stanford Humanities Center, n.d.).

7.5.4 Dr. Katalin Karikó—Winner of Nobel Prize in Physiology or Medicine, 2023

As a 2023 winner of Nobel prize Dr. Katalin Karikó along with Dr. Drew Weissman for their groundbreaking work in creating messenger RNA (mRNA) technology that developed vaccines used for COVID-19 during the pandemic, such as Pfizer-BioNTech and Moderna (Nobel Prize, 2023), she was an embodiment of diligence. Karikó worked for over 30 years on mRNA, and along the way faced refusal and rejection by institutions to obtain research permissions and funding. Karikó received international acclaim during the COVID-19 pandemic as over 13 billion doses of mRNA vaccines have been given around the world as of 2023, and the technology is being applied toward cancer immunotherapy, HIV, influenza, and autoimmune diseases (World Health Organization, 2023). Dr. Karikó has received dozens of international patents and has been a Senior Vice President at BioNTech, instrumental in working with mRNA with real-world applications.

These women exemplify that scientific advancements are not just products of intelligence and experimentation, but also require resilience against gender-based challenges. They have changed industries, saved lives, and pushed boundaries to create a more equitable and sustainable scientific ecosystem.

7.6 WOMEN LEADERS IN CLIMATE INNOVATION AND THE GREEN ECONOMY

As climate change accelerates and more aggressive federal and state policies respond to urgent issues that governments, businesses, and organizations face, we are witnessing that the US green economy is undergoing a major shift in leadership. Women are becoming prominent players in shaping this transition by working on climate innovation developments, green job creation, and sustainable business growth. Although women leaders in green and brown jobs are increasing in visibility and representation, their contributions are still not well represented in mainstream narratives about environmental innovations. Here, we will offer a review of women leaders in climate innovation, the policies they are capitalizing on, and the empirical evidence that supports the connection between gender diverse leadership and better environmental outcomes.

7.6.1 The US Green Policy Landscape and Gendered Opportunities

The prime federal investment in climate resilience and clean energy in American history came about with Inflation Reduction Act (IRA), was passed in 2022, and under which $369 billion was allocated to climate-focused and energy-focused projects (U.S. Department of Energy, 2023). The IRA created an ecosystem to support women entrepreneurs and environmental justice organizations. Additionally, the "Office of Economic Impact and Diversity" located within the Department of Energy has made inclusive participation in energy-related fields a key operational value, and more women are leading clean tech companies in leadership roles (U.S. Department of Energy, Office of Economic Impact and Diversity, 2023).

7.6.2 Case Studies: Women Leading Climate Innovation in the USA

With a climate crisis reaching unprecedented levels, women in America have become innovative leaders in the realm of the environment. Their efforts are advancing solutions to environmental challenges and embedding sustainability in public policy, business, and community initiatives. The case studies below illustrate pioneering women who are redefining leadership in the green economy.

7.6.2.1 Lisa P. Lisa Jackson—Apple, Former EPA Administrator

Lisa Jackson was appointed as an Administrator in the Environmental Protection Agency (EPA) under the Presidency of Barack Obama, and currently leads Apple's sustainability and social initiatives. At Apple, she has overseen the transition to 100% renewable energy for all operations and the use of recycled materials within their major product lines (Apple ESG Report, 2023). Lisa's simultaneous influence as a government official and in the private sector reflects how women leaders span the regulatory and entrepreneurial realms in support of climate innovation.

7.6.2.2 Revathi Advaithi—CEO, Flex (San José, CA)

Revathi Advaithi, an expatriate of Indian nationality, has been the CEO of Flex Ltd., a Fortune Global 500 company with headquarters in San José, California, since 2019. Under her leadership, Flex adopted a target to reduce CO2 emissions by 2040 with an interim target of 50% reduction in its operational carbon emissions by 2030 (Flex, 2023). Her commitment to environmental, social, and governance (ESG) initiatives has helped the company gain recognition on the CDP Climate Change "A List" for its commitment to climate action. Advaithi is also a noted champion of workplace diversity and inclusion and has increased women in the executive ranks from 18% to 27% between 2019 and 2023 (TIME, 2023). Advaithi also actively participates in global sustainability efforts by contributing to thought leadership on systemic climate resilience, circular economy practices, and responsible industrial innovation (Advaithi, 2024). The notable achievements of Advaithi demonstrate the rising impact of women expatriates in leading corporate climate policy for manufacturers and technology organizations in the USA.

7.6.2.3 Marilyn Waite—Clean Energy Finance as a Service, Climate Finance Fund

Marilyn Waite is an American expatriate who is a globally competent clean energy finance expert and serves as the Managing Director of Climate Finance Fund (CFF). Waite is most well-known for her work on blended finance mechanisms aimed at galvanizing private and philanthropic investment into clean energy transition (Climate Finance Fund, 2023). Her work experience includes leadership roles with AREVA, Project Drawdown, and Village Capital, focused on climate technology innovation and energy access. Waite is also the Director of Climate First Bank, where she highlights equitable access to climate finance (Trellis, 2024). Waite has been published in many forums including her own publication Sustainability at Work as well as Financial Times and GreenBiz, helping to create a formidable platform to leverage inclusive, data-informed approaches to clean energy transformation (Waite, 2021).

7.6.3 Research Evidence: Gender and Environmental Innovation

Moreover, there is developing evidence base pointing to the role of gender in leadership contributing to better ESG outcomes. Atif et al. (2023) found that companies with women leaders implemented proactive climate strategies. Similarly, Sandretto et al. (2025) found that environmental disclosures for US companies led by female executives were scored 27.6% higher and received fewer regulatory penalties concerning environmental violations.

7.6.4 Policy Implications

Women leaders in green enterprises are changing the expression of sustainability. In their various forms of leadership, they are contributing to ecosystem building around federal policies, venture capital, startup activity, and corporate ESG transformations. Women leaders have the capacity to institutionalize mentorship, access to capital, and gender-equity benchmarks in US federal programs for decades to come. Many arenas in future research may study the longitudinal impact of women-led climate actions on the competitiveness of the USA in global sustainability transitions.

7.7 WOMEN IN ARTIFICIAL INTELLIGENCE AND EMERGING TECH LEADERSHIP

Diffusion of AI, quantum computing, and blockchain in the USA is shaping economic competitiveness while also prompting troubling questions related to inclusion and equity. Although women are actively engaged in innovation, they continue to lack representation in these emerging technologies. According to the World Economic Forum (2023), just 22% of AI professionals in the USA are women. This section is about American women's barriers and breakthroughs in AI and emerging technologies, and highlights their leadership, innovations, and advocacy around ethical technology.

7.7.1 Case Studies: US Women Leading AI and Ethical Tech

Amidst the changing future of economies and societies through AI, more and more American women are innovating in the domains of technology and ethics. These women leaders are driving AI forward, but they are also giving voice to fairness, accountability, and inclusivity within digital systems. The following case studies highlight women who are creating the ethical frameworks of technologies that are emerging within the USA.

7.7.1.1 Joy Buolamwini—Entrepreneur, Algorithmic Justice League

As a graduate researcher at MIT Media Lab, Joy Buolamwini gained global attention from her "Gender Shades" study, in which she identified racial and gender bias in commercial face recognition systems (Buolamwini & Gebru, 2018). Buolamwini's work has been cited in important hearings in congress, and she also called for more transparency and regulation of existing AI systems.

7.7.1.2 Fei-Fei Li—Co-Director, Stanford Human-Centered AI Institute

Fei-Fei Li, as a former Chief Scientist of AI at Google Cloud, advocated for responsible AI development and data collection that promote inclusion. At Stanford, Li led a group of colleagues to upscale women representation in AI research from the ground up, founding a national nonprofit called AI4ALL, which has reached thousands of students (Deng et al., 2009).

7.7.1.3 Latanya Sweeney—Professor, Harvard University and Technologist, FTC

Sweeney was an early contributor to the areas of data privacy and algorithmic fairness. She demonstrated that online ad delivery systems use racial disfavor to unfairly disseminate information, and credit scoring algorithms display significant bias. Sweeney joined the "U.S. Federal Trade Commission" as a distinguished Chief Technologist. Collaborating with lawmakers, she strived to contribute to creating regulatory frameworks for algorithmic transparency and accountability (Sweeney, 2013).

7.7.2 Institutional Gaps and Emerging Women Leadership in AI Ecosystems

While gender equity in AI and emergent skills has become increasingly visible in America, women's leadership paths continue to be obstructed by structural and institutional barriers. Gaps remain across multiple domains, including access to venture capital and technical representation, along with executive decision-making structures. According to the AI Now Institute, women represent a mere 15% of AI faculty at major US universities and less than 18% of authors at top-tier AI conferences (West et al., 2019). Moreover, women hold only 13% of executive positions in machine learning groups in large tech companies and only 20% of AI startups financed by venture capital firms have female founders (Zang et al., 2021). These numbers are worsened by algorithmic bias and gendered exclusions in R&D ecosystems. Nonetheless, a counter-narrative is emerging: AI startups founded by women generated 35% higher ROI than those founded by men (Accenture Research, 2022), indicating unrealized opportunities for innovation. Systemic responses to these inequities are needed, such as gender-aware funding models, inclusive accelerator programs, and equity-oriented public policy. American women leaders, while underrepresented, are key to the development of ethical, inclusive, and privacy-sensitive AI architectures, to bridge the gender divide in digital technology (Table 7.2).

TABLE 7.2
Global Women Innovators Driving Sectoral Transformation

Name	Sector	Contribution	Outcome
Indra Nooyi	FMCG	Purpose-driven leadership	Greener portfolios
Katalin KarikÃ³	Healthcare/Biotech	mRNA development	Nobel Prize, global vaccination
Joy Buolamwini	AI Ethics	Algorithmic bias activism	US legislative impact
Lisa Jackson	Corporate Sustainability	Apple's climate strategy	Net-zero milestones

Source: Compiled from publicly available profiles and impact reports (e.g., Nooyi, 2021; Apple Environmental Progress Report, 2023; BioNTech ESG Report, 2024).

7.8 WOMEN LEADERS SHAPING INNOVATION POLICY AND POLICY THINK TANKS IN THE USA

While the majority of the discussion around women and innovation seems to take place in the private sector or entrepreneurial contexts, equally important but less visible area is public policy. Women in the USA are now taking notable leadership roles in making national innovation policy, science advisory boards, and policy think tanks. They are impacting how regulations are made, making sure that technology funding includes underrepresented groups, and deciding what research agendas focus on for the future and to maintain competition.

7.8.1 Arati Prabhakar—Director, White House Office of Science and Technology Policy (OSTP)

She is the first woman and first person of color to fill this role in the history of the country (Prabhakar, 2023). She reports to the President of the USA as his Chief Science Advisor and must provide advice on AI regulatory issues, semiconductor investments through parity passive means (e.g., CHIPS and Science Act), and biotechnology innovation (National Science and Technology Council, 2023). As the Chief Science Advisor to the President, Dr. Prabhakar has directed efforts in ensuring federal research and development investments consider diverse perspectives and inclusive outcomes (OSTP, 2023).

7.8.2 Frances Colón—Science Diplomat and Innovation Policy Advisor

Dr. Frances Colón, an American science diplomat born in Puerto Rico, has been a major influence in developing US science and environmental diplomacy (Colón, 2015). From 2008 to 2017, she was the "Deputy Science and Technology Adviser to the Secretary of State," helping to establish the Energy and Climate Partnership of the Americas, embedding science-based climate diplomacy into US foreign relations (West, 2020). She also worked on building the innovation ecosystem in Latin America and leveraging cross-border collaborations in the clean energy transition (Colón, 2015). Currently, Dr. Colón serves on the "President's Council of Advisors on Science and Technology (PCAST)" to elevate federal strategic priorities in climate resilience, advancing equity in STEM, and assessing global innovation policy. Through her hemispheric influence, she is illustrative of how women expatriates shape inclusive and science-based policy design for innovation and governance between diplomacy and innovation (PCAST, 2024).

7.9 MENTORSHIP, NETWORKS, AND INSTITUTIONAL ECOSYSTEMS: A WASHINGTON PERSPECTIVE ON WOMEN INNOVATORS IN THE USA

Certainly, the hurdles of gaining access to capital and legislative reform are both adequately considered concerns for women's development in innovation ecosystems, and growing empirical literature is highlighting the impending importance of mentorship networks and institutional enabling as additional critical elements. There are various organizations across the USA that are working to provide more inclusive spaces for young women to explore, create, and grow innovations, which include nonprofits, accelerators, and academic programs (Gonzalez & Clark 2022). In particular, girls from underrepresented communities are being exposed to coding and AI for the first time in initiatives such as Girls Who Code and AI4ALL. Girls Who Code is proud of their 300,000+ alumni, many of whom are leaders in their startups or federal science agencies (Girls Who Code, 2023). AI4ALL was started by AI leader Fei-Fei Li and runs research labs at universities with the goal of increasing diversity in AI graduate programs, and they have been successful in showing an increase in diversity (Li, 2024). The same increased venture success has been shown in focused accelerators such as Techstars Female Founders Catalyst and Women Who Tech (Corn, 2023; Crunchbase, 2023), where participating women are more successful than their male peers in securing funding in the general startup ecosystem (2.3 times the likelihood of success).

In addition to all the programs that support startup incubation, the role of continued mentorship while in the workforce is vital in supporting innovation-related outcomes. The "National Center for Women & Information Technology (NCWIT)," which promotes equity for women in computing through institutionalized solutions, like unconscious bias training, peer-based communities, and fellowships, saw an increase of 17% (NCWIT, 2023) in workforce retention for engaged and active women in technical roles. A survey by Accenture (2022) found that when US companies had formal woman mentorship programs, they saw a 46% increase in female patent filings, and overall teams had a 29% greater contribution to innovation scores. Further, the National Science Foundation (NSF) found that women engaged in mentorship networks were 63% more likely to continue ventures long-term beyond 3 years (NSF, 2023). These results highlight that mentorship is not implementing an additional intervention but the necessary infrastructure to develop a sustainable and high-performing pipeline of women innovators. Inclusive ecosystems are more than an ethical imperative; they show measurable benefits in terms of innovation performance.

7.10 CASE OF WOMEN INNOVATORS IN POLITICS

America has long been a worldwide leader for democratic institutions and economic power, no woman has ever stood for election as "President of the United States". The first woman to run for US president was Victoria Woodhull in 1872, who ran on a progressive agenda of women's suffrage, labor, and social justice at a time when women had no legal right to vote (Sanbonmatsu, 2019; Lawless & Fox, 2020; Center for American Women and Politics, 2024). Over a hundred years later, Kamala Harris became the most recent woman to run for US president in the 2024 election cycle (Dittmar, 2020; Smooth, 2021). This instance and Lopez's continued quest for female representation in executive roles reveal the stark variability in women serving at the highest levels of executive leadership. To date, there have only been four women to receive major-party presidential nominations, or achieve serious candidacy status, and no women in the USA have been elected president, while several other countries have elected women heads of state (Lawless & Fox, 2020; CAWP, 2024).

Women have also made substantial progress in legislative and executive bodies, having a truly transformational effect on federal and state-level policies. In 2024, women held 125 seats in the US House of Representatives (28.7%), 25 seats in the US Senate (25%), and 12 women served as governors in US states (Lawless & Fox, 2020; Center for American Women and Politics, 2024). The following cases present how women leaders actively not only challenged patriarchal systems of power but also employed equitable and inclusive policies in health, economic, and civil rights spheres.

7.10.1 Elaine L. Chao—US Cabinet Secretary

Elaine L. Chao, originally from Taiwan, immigrated to America in her younger years. She is now one of the most influential women in the history of the US federal government with Asian roots. In fact, Chao was the first Asian American woman appointed to the Cabinet of a US president. Her distinguished service includes being Secretary of Labor from 2001 to 2009 under President George W. Bush, and Secretary of Transportation from 2017 to 2021 under President Donald Trump, during which she became the longest-serving cabinet member since World War II. The journey of Chao from corporate executive to high-level federal service reflects leadership developed from an immigrant background. Her presence in federal government has been particularly felt in policy areas like modernization of US infrastructure, workforce development, and community resiliency (McClain & Stewart, 2019; Dittmar, 2020).

7.10.2 Stephanie Murphy—US Representative

Stephanie Murphy, from Vietnam, was very young when her family moved to the USA as refugees in 1979. Beginning as a national security professional, she attended the College of William & Mary and Georgetown University (Georgetown University, 2021), transitioning into electoral politics. After her first election in 2016, Stephane Murphy became the first Vietnamese American woman voted into the US Congress as a member of the 7th Congressional District of FL representing from 2017 through 2023 (U.S. House of Representatives, n.d.). Known for her bipartisan teamwork and national security experience, she was appointed to serve on the House Select Committee on the January 6th Attack (Thompson, 2022). Her appointment reinforced her position in developing policy during a politically turbulent time in the US Murphy's election and service in Congress, demonstrating the rise of immigrant women into federal leadership roles and their impact on resilience, oversight and inclusive governance (Biographical Directory of the United States Congress, n.d.).

7.10.3 Kamala Harris—First Woman Vice President of the USA

Kamala Harris made history by becoming the first female, first women of minority, and first American Vice President of the United States in 2021 with South Asian origins. Prior to becoming Vice President, Harris was elected "District Attorney of San Francisco," then "Attorney General of California," known for her work on criminal justice reform, consumer rights, and data privacy (White House, 2023).

She played a major role in the passage of the Infrastructure Investment and Jobs Act (2021) as the Vice President and is leading national efforts on immigration reform, reproductive rights, and racial equity (White House, 2023). Her candidacy for the presidency in 2024 also indicated a major moment for intersectional representation in US politics.

These women illustrate the slow and overdue change in American political leadership. Their lives and careers illustrate how substantive representation (in which leaders genuinely shape policy for underrepresented communities) can emerge from systems where descriptive representation (number) has been curtailed. As political scholars intimate, women in executive and legislative roles are correlatively associated with increasing attention to social equity, education, reproductive rights, and public health, thus increasing the institutional responsiveness of legislatures in representing marginalized constituencies (Dittmar, 2020; Lawless & Fox, 2020).

7.11 LEADERSHIP PARADIGMS AND THE STRATEGIC DISTINCTION OF WOMEN LEADERS

Traditional leadership models based on transactional power and productivity (Burns, 1978; Bass, 1990) are steadily being rendered inadequate in post-industrial knowledge economies, driven by diversity, inclusion, and decentering power in organizational eco-systems. A push toward transformational, emotional, and participative leadership (Goleman, 2000; Avolio & Gardner, 2005) has increased since COVID-19 to address workers' current priorities, including psychological safety, ethical stewardship, and flexibility (Edmondson, 2019; Gartner, 2021). Meta-analysis demonstrates that women outperform men on the most essential leadership traits such as integrity, resilience, and inspirational leadership (Zenger & Folkman, 2020). Relationship-based leadership approaches that rely on trust, transparency, and inclusivity also positively influence variables such as organizational citizenship behaviors and engaged employees (Eagly & Carli, 2003; Nishii, 2013). Foundational to these values is the tendency for women managers to check in with employee wellness more frequently (McKinsey & Company & LeanIn.Org, 2021; Sandretto et al., 2025).

Statistically, women's leadership is significantly associated with higher ESG deficits, as women-led firms (e.g., women-led public companies) score better: 27.6% higher ESG disclosures and lower propensity to be fined for violations (Sandretto et al., 2025). Studies demonstrate a positive relationship between board gender diversity and corporate social responsibility (Post & Byron, 2015). Notably, leaders with a feminine perspective bring long-term risk-oriented thinking and a sustainability focus to their decision-making (Barber & Odean, 2001; Adams & Funk, 2012)—and these decision traits produce real material economic returns: gender-diverse firms represent 34% higher shareholder returns (Catalyst, 2022). These shifts demonstrate the importance of enacting inclusive leadership pathways through policies and structures to allow organization members to feel safe and respected, with data related to gender diverse leaders link to improvements in resilience, profits, and sustainable growth (Jasmin, 2017; Sandretto et al., 2025) (Table 7.3).

TABLE 7.3
Prevalence of Leadership Styles Among Women Leaders

Leadership Style	Prevalence Among Women Leaders (%)	Source
Democratic/Participative (Corporate Sector)	67	Alqahtani (2020)
Democratic/Participative (Government Sector)	57	Alqahtani (2020)
Transformational	~60	Eagly (2005)
Authentic	~70	Lucey and Burke (2022)
Servant Leadership	~65	Nguyen et al. (2024)

Source: Author's own construction based on cited studies.

7.12 BREAKING DOWN STRUCTURAL BARRIERS: WOMEN LEADING AND INNOVATING IN MODERN AMERICA

The accounts of women in leadership highlight a recurring theme—an ongoing struggle against entrenched social, institutional, and cultural barriers. While there has been progress, structural inequities still exist to hamper the leadership trajectory of women, especially in male-dominated occupations such as technology, finance, and policy. Ongoing academics and policy-level dialogues are necessary to dismantle these structural barriers and create an inclusive innovation ecosystem (Eagly & Heilman, 2016; Madsen & Andrade, 2018).

7.12.1 Gender Bias and Leadership Stereotypes

Numerous studies have documented that gender stereotypes consistently connect leadership attributes, like assertiveness, decisiveness, and strategic thinking, with men, while women are seen as more communal and nurturing (Eagly & Carli, 2007; Koenig et al., 2011). As of 2023, women are holding only 10.4% of the Fortune 500 CEO positions in America, while 47.7% of the labor force constitutes women (Catalyst, 2023). This illustrates the problem of structural bias in the leadership pipelines.

7.12.2 Lack of Role Models and Mentorship

The scarcity of women in senior leadership roles, especially in STEM or high-growth industries, further creates an ongoing cycle of invisibility. A longitudinal study by Xu (2008) showed underrepresented female faculty and female leaders in engineering departments were less likely to mentor junior women primarily because of their lack of representation. The lack of visible role models contributed to feeling prohibited for younger women from seeking careers in these streams (Dasgupta & Stout, 2014). According to evidence, companies led by women are more likely to advocate for women and thus advance a multiplier effect (Latu et al., 2013).

7.12.3 Discrimination and Pay Inequality

While there are legal mechanisms to ensure equal pay, women are still at a disadvantage. According to the U.S. Census Bureau (2022), women earn about 82 cents for every dollar paid to men. The wage gap is stark in case of minority women. Experiments show that the same resumes are rated as lower quality when they include a historically female name (Moss-Racusin et al., 2012). Cases of occupation segregation and limited access to higher-paying jobs underlie inequities that persist today (National Science Foundation, 2007).

7.12.4 Access to Capital and Funding Gaps

Access to venture capital remains highly inequitable. Women pioneering entrepreneurial ventures attained only 2.1% of total venture capital funding in the USA in 2022 (PitchBook, 2023). The gap reflects perceptions of gender from capital providers and not the competency of women's ventures. Kanze et al. (2018) found that women entrepreneurs face more preventive inquiries, such as risk-averse questions, whereas male entrepreneurs are asked more opportunity questions. This difference, in effect, inscribes bias when individuals assess investment opportunity.

7.12.5 Work-Life Balance and Social Prospects

Work-life conflict becomes a barrier for women moving into influential leadership. Pew Research Center (2015) shows that 50% of working mothers reported that parenting made it more difficult to advance their careers compared to only 39% of fathers. The imbalanced expectations derive from patriarchal norms, creating an uneven caregiving burden for women and the ability for many high-achieving women to leave (or delay) demanding occupations in environments without supportive institutional practices like flexible hours of work and workplace child care (Hill et al., 2001).

7.12.6 Cultural Hostility and Microaggressions

Cultures in male-dominated industries often reflect implicit bias and an unfriendly environment. According to Sue et al. (2007), women routinely endure microaggressions, being cut off, their ideas being ignored, or being confused for those with less authority. Research from McKinsey & Company (2020), in their "Women in the Workplace" report, found that one in three women deals regularly with such behaviors, contributing to higher attrition and less engagement.

7.12.7 Underrepresentation in Decision-Making and Policy Roles

The "glass ceiling" remains pervasive, especially in influential industries affecting future innovation. Women occupy only 28% of senior management globally, and, at the lowest levels of representation in finance and tech, drop below 20%

(World Economic Forum, 2023). Bloch et al. (2021) maintained that with greater inclusivity is Associated with greater performance, but they also underscore the need for systemic reform for removing stereotypes and structural gate-keeping. Advancing the number of women in boardrooms is correlated to declarations of more inclusive corporate policies, improved ESG performance, and better organizational resilience (Terjesen et al., 2009).

7.12.7.1 Underrepresentation in Decision-Making Roles

Only when we see an equitable representation of women in these roles, that is when onset of transformation will happen. This will lead to policies that are more inclusive in nature. The "glass ceiling" continues to be an issue, especially in tech and finance (Figure 7.1).

7.12.7.2 Gender Representation and the Structural Barriers of Leadership in the USA

Despite improvements in education and workforce engagement, women remain severely underrepresented in top leadership positions. In 2023, women comprise only 30.4% of all board seats in S&P 500 organizations, and 27.9% of Congress, with the presidency still unoccupied by a woman (CAWP, 2023; Spencer Stuart, 2023). This underrepresentation has an impact on governance. For example, boards with gender diversity are more likely to promote well-being for employees and corporate social responsibility (Post & Byron, 2015), and female legislators prioritize health care, education, and family policy at noticeably higher rates

Source: Author's own adaptation based on data from Pew Research Center (2023). Pew Research Centre

FIGURE 7.1 Percentage of women CEOs in Fortune 500 companies. Source: Author's own adaptation based on data from Pew Research Center (2015). Pew Research Centre.

(Schwindt-Bayer, 2010). Women also experience a "double bind" in that they can be penalized when they display either masculine or feminine leadership characteristics (Catalyst, 2022).

Maternity and care responsibilities also contribute to women's exits from many leadership channels. The USA is the only OECD nation without a federally mandated paid maternity leave policy, which leads many women to leave the workforce after childbirth (OECD, 2022). The penalty of motherhood means that wages will decrease, and the opportunity for advancement will become more limited (Budig & England, 2001). Dedicated organizations that address the structural constraints of institutional supports, such as rigid hours, lack of childcare, or weak re-entry, contribute to the gender gap. Research shows care represents a 30% higher likelihood for women transitioning to part-time or lower-skilled jobs (ILO, 2020). While returnship programs and flexible work can decrease the likelihood of care-based exits, Americans have not adopted these approaches easily into US corporations (Kossek et al., 2006).

7.13 US GOVERNMENT AND PRIVATE SECTOR INITIATIVES SUPPORTING WOMEN INNOVATORS

In recent years, US government agencies and private organizations have launched specific efforts to increase gender equity in the innovation economy. The US Department of Energy's Women in Clean Energy initiative provides fellows and training in leadership for women involved in climate and energy innovation (DOE, 2022). The NSF ADVANCE program similarly seeks an increase in the numbers and advancement of women faculty in academic STEM careers with an emphasis on systemic institutional advancement (NSF, 2023). The private sector support has varied widely as well. The Techstars Female Founders Fund provides funding and access to investor networks to early-stage women-led business ventures (Techstars, 2022). The Women Who Tech Challenge provides prizes for competitions around global events that showcase and provide funding to women-led tech innovations (Women Who Tech, 2022).

While the various programs provide meaningful opportunities toward gender equitable access to the innovation economy, there are limitations that have come under review. For instance, while government programs, such as Women in Clean Energy, provide a cohort experience for fellows, rural outreach is limited, and there is no long-term support for trainees who would like to integrate their program experience into job or post-grad experiences. Similarly, NSF ADVANCE has had more institutional success in bringing more women into the academy than helping women to bridge out and transition to commercial STEM industries. The private sector initiatives above are heavily weighted toward providing a portion of seed funding, not necessarily toward long-term scaling support for women entrepreneurs. Furthermore, more competitive and episodic formats support the exclusion of high-quality and high-potential opportunities due to eligibility criteria and a pattern of licensed qualification. For these programs to create a sustainable impact, inclusive program design, funding sustainability, and outreach with an intersectional lens are necessary (Tolan et al., 2023).

7.14 STRUCTURAL FOUNDATIONS FOR TRANSFORMATIVE WOMEN'S LEADERSHIP

Although women's numerical leadership representation is a crucial first step, it does not guarantee meaningful leadership outcomes. Normative equality in leadership must support institutional contexts that foster coalition-building, informal influence, and access to decision-making power. Meaningful change occurs when organizational structures support gender equity beyond policies or rhetoric, and women are able to unify themselves around collective agendas and create strategic impact (Domingo et al., 2015). The case studies of US entrepreneurship, activism, and public policy documentation illustrate how women leverage their lived experiences to engender systemic innovation when positioned in enabling institutions.

That women's leadership will lead to change in its conceptual programmatic form is less likely in situations without feminist institutional theory's abnegation of primitive yet resistant informal power relations such as patronage, elitism, and cultural bias (Chappell & Waylen, 2013). Without structural transformation, women's individual success stories risk being treated as symbolic rather than catalytic. In fact, Woetzel et al. (2015) provide evidence that organizations with diverse leadership demonstrate both enhanced capacity for innovation, civic responsiveness, and greater long-term economic performance. Therefore, supporting women's leadership is not deficiently responsible—investing in women's leadership is the ethically responsible thing to do, but it is also the economically and socially smart thing to do for sustainability and inclusive growth (Table 7.4).

TABLE 7.4
Major US-linked Initiatives Supporting Women in Innovation and Leadership

Initiative	Administering Body	Focus Area	Impact	Criticism/Shortcoming
Women in Clean Energy	US Department of Energy	Climate Tech	Training and fellowships	Limited rural outreach; lacks follow-up career support mechanisms
NSF ADVANCE	NSF	Academic STEM	Faculty gender equity	Predominantly benefits research universities; limited industry linkage
Techstars Female Founders Fund	Private VC	Startup Capital	Seed funding support	Focuses on early-stage; less support for post-seed growth
Women Who Tech Challenge	NGO	Innovation Competitions	Exposure and mentorship	Competitive model limits access; awards few winners relative to demand
Women's Global Development and Prosperity (W-GDP) Initiative	US Government (USAID, Department of State)	Global Women's Leadership Development	Leadership training, entrepreneurship funding in developing regions	Primarily supports foreign-based leadership growth; limited domestic scale

Source: Author's compilation based on official agency and program documents (Techstars, 2022; NSF, 2023; USAID, 2023).

7.15 WOMEN'S EMPOWERMENT AS A PATHWAY FOR THE FUTURE

This study indicates the undeniable power for change women leaders can provide across industries. Female representation will not only drive gender equity, but they also serve an important role in advancing best practices, ethical governance, and inclusive policies. There are structural changes necessary, such as mentorship networks, gender responsive investment strategies, inclusive parental leave, and anti-bias training, to facilitate the dismantling of barriers. The importance of supporting women is not merely symbolic, but a structural necessity for future sustainable socio-economic projections (UN Women, 2023; OECD, 2024). Adopting, supporting, and institutionalizing women's leadership is vital if the next generation of innovation is to be genuinely representative of the realities of society.

7.16 CONCLUSION

Historically, women's contributions to the innovation economy in the USA have been prominent, but their leadership has often been overlooked or minimized. As current-day innovators, women often dismantle older systems of operation, create ethical, equitable, and collaborative spaces, and work to radically transform systems and practices across science, technology, policy, and enterprise (Fisher et al., 2015). In bringing women's leadership and innovation together, this chapter traced the historical evolution of women's leadership, as highlighted by key legal milestones such as the passing of the 19th Amendment (1920) and the equal pay legislation in the Equal Pay Act (1963), mapped women's leadership within today's high-value sectors (even as we recognized persistent challenges that persist—gender bias, limited representation, and underrepresentation particularly in STEM, and women's limited access to venture capital).

The case studies included herein, spanning the new purpose-led corporate design of Indra Nooyi, or Katalin Karikó—who won the Nobel Prize for her pioneering breakthroughs in mRNA—evidence how women leaders continue to reimagine and evolve the narrative of innovation in contemporary society. The above illustrates how women leaders are not only role models in their own right, but also designing inclusive larger institutional ecosystems. Studies have shown organizations with gender-diverse leadership teams are more innovative, resilient, and competitive (Hunt et al., 2020). However, systemic barriers continue to exist: boardrooms lack gender parity, disparities in earnings persist, structural supports for child care and caregiving has singularly gone vertical (Rhode, 2017). In order to support transformative innovation, pathways for gender-inclusive leadership must be embedded in everyday institutional processes related to policy, funding, and organization design. Without the engagement of women at all levels of equality, national and corporate sustainability and innovation agendas will ultimately remain incomplete.

BIBLIOGRAPHY

Accenture. (2021). Pipeline research: Closing the gender equity gap in AI leads to higher revenue. Accenture Research White Paper

Accenture. (2022). The equity effect: How mentorship moves innovation forward. https://www.accenture.com/us-en/about/inclusion-diversity/equity-effect

Adams, R. B., & Funk, P. (2012). Beyond the glass ceiling: Does gender matter? *Management Science*, 58(2), 219–235. https://doi.org/10.1287/mnsc.1110.1452

Advaithi, R. (2024). Revathi Advaithi: Agenda contributor profile. World Economic Forum. https://www.weforum.org/stories/authors/revathi-advaithi/

AI4ALL. (2023). Increasing diversity in AI through education and mentorship: 2023 impact report. https://ai-4-all.org/impact/

Alqahtani, T. H. (2020). The status of women in leadership. *Archives of Business Research*, 8(3), 294–299.

American Nurses Association. (2023). 2023 nursing trends and salary survey results. *American Nurse Journal*. https://www.myamericannurse.com/2023-nursing-trends-and-salary-survey-results/

Apple Inc. (2023). Environmental progress report 2023. https://www.apple.com/environment/pdf/Apple_Environmental_Progress_Report_2023.pdf

Atif, M., Hossain, M., Alam, M. S., & Goergen, M. (2023). Board gender diversity and climate change innovation: Global evidence. *Journal of Innovation & Knowledge*, 8(3), 100374.

Avolio, B. J., & Gardner, W. L. (2005). Authentic leadership development: Getting to the root of positive forms of leadership. *The Leadership Quarterly*, 16(3), 315–338. https://doi.org/10.1016/j.leaqua.2005.03.001

Barber, B. M., & Odean, T. (2001). Boys will be boys: Gender, overconfidence, and common stock investment. *The Quarterly Journal of Economics*, 116(1), 261–292. https://doi.org/10.1162/003355301556400

Bass, B. M. (1990). *Bass & Stogdill's Handbook of Leadership: Theory, Research, and Managerial Applications* (3rd ed.). Free Press.

Biographical Directory of the United States Congress. (n.d.). Stephanie murphy. https://bioguideretro.congress.gov/Home/MemberDetails?memIndex=M001204

BioNTech. (2024). BioNTech environmental, social, and governance (ESG) report. https://www.biontech.com/int/en/sustainability/reporting.html

Bloch, F., et al. (2021). Breaking barriers: The impact of co-leadership on the gender gap in leadership participation. Social Science Research Network. https://papers.ssrn.com/sol3/papers.cfm?abstract_id=5054412

Bosma, N., Kelley, D., & Amorós, J. E. (2023). *Global Entrepreneurship Monitor 2023/24 Women's Entrepreneurship Report*. Global Entrepreneurship Research Association.

Brush, C. G., Greene, P. G., Balachandra, L., & Davis, A. E. (2020). The gender gap in venture capital- progress, problems, and perspectives. *Venture Capital*, 22(2), 115–136. https://doi.org/10.1080/13691066.2017.1349266

Budig, M. J., & England, P. (2001). The wage penalty for motherhood. *American Sociological Review*, 66(2), 204–225. https://doi.org/10.1177/000312240106600203

Buechler, S. M. (1990). *Women's Movements in the United States: Woman Suffrage, Equal Rights, and Beyond*. Rutgers University Press.

Buolamwini, J., & Gebru, T. (2018). Gender shades: Intersectional accuracy disparities in commercial gender classification. *Proceedings of Machine Learning Research*, 81, 1–15

Burns, J. M. (1978). *Leadership*. Harper & Row.

Burns, U. (2021). *Where You Are Is Not Who You Are: A Memoir*. HarperCollins.

BusinessWire. (2024, March 7). PitchBook reports show female founded companies secured record high proportion of deal value in 2023. https://www.businesswire.com/news/home/20240306750545/en/PitchBook-Reports-Show-Female-Founded-Companies-Secured-Record-High-Proportion-of-Deal-Value-In-2023

Cabello-Hutt, T., Caballero, V., & Sosa, L. (2020). Changes in work and care trajectories during the transition to motherhood: Evidence from Chile. *Social Science Research*, 85, 102371. https://doi.org/10.1016/j.ssresearch.2019.102371

Catalyst. (2007). The double-bind dilemma for women in leadership: Damned if you do, doomed if you don't. Catalyst. https://www.catalyst.org/research/the-double-bind-dilemma-for-women-in-leadership-damned-if-you-do-doomed-if-you-dont/3

Catalyst. (2022). Why diversity and inclusion matter: Financial performance. https://www.catalyst.org/research/why-diversity-and-inclusion-matter-financial-performance/

Catalyst. (2023). Women CEOs of the S&P 500. Catalyst. Retrieved August 16, 2024. https://www.catalyst.org

Center for American Women and Politics. (2024). Women in elective office 2024. Rutgers University. https://cawp.rutgers.edu/facts/levels-office/current-numbers

Center for American Women and Politics (CAWP). (2023). New records for women in the U.S. congress and house. Rutgers University, Eagleton Institute of Politics. https://cawp.rutgers.edu/news-media/press-releases/new-records-women-us-congress-and-house

Centers for Disease Control and Prevention. (2022). COVID data tracker: Vaccinations in the United States. https://covid.cdc.gov/covid-data-tracker/#vaccinations

Chappell, L., & Waylen, G. (2013). Gender and the hidden life of institutions. *Public Administration*, 91(3), 599–615. https://doi.org/10.1111/j.1467-9299.2012.02104.x

Climate Finance Fund. (2023). About us – Mobilizing capital for climate solutions. https://climatefinance.fund/

Colón, F. A. (2015, September 11). Deputy science and technology adviser to the secretary of state: Science diplomacy and innovation. Cell Therapy Conference Profiles. https://web.archive.org/web/20210123150236/http://celltherapyconference2016.com/2015/09/11/dr-frances-colon/7

Corn, A. (2023, May 18). Women get 2.3x more returns in equity-focused accelerator programs. Forbes. https://www.forbes.com/sites/allisoncorn/2023/05/18/women-accelerators-return-data/

Crunchbase. (2023). Women in venture 2023 update: Funding gaps and success stories. https://news.crunchbase.com/diversity/women-in-venture-2023/

Dasgupta, N., & Stout, J. G. (2014). Girls and women in science, technology, engineering, and mathematics: STEMing the tide and broadening participation in STEM careers. *Policy Insights from the Behavioral and Brain Sciences*, 1(1), 21–29. https://doi.org/10.1177/2372732214549471

Deng, J., Dong, W., Socher, R., Li, L.-J., Li, K., & Fei-Fei, L. (2009). ImageNet: A large-scale hierarchical image database. *Proceedings of the IEEE Conference on Computer Vision and Pattern Recognition (CVPR)*, 30, 248–255.

Department of Energy. (2022). *Clean Energy Education & Empowerment (C3E) Initiative*. U.S. Department of Energy. https://www.energy.gov/eere/wipo/clean-energy-education-empowerment-c3e-initiative

Deutschland.de. (2021, March 8). The woman behind the vaccine. https://www.deutschland.de/en/topic/knowledge/coronavirus-vaccine-ozlem-tureci-of-biontech

Dittmar, K. (2020). *Representing women: Women in the U.S. Congress*. Oxford University Press.

Domingo, P., Holmes, R., O'Neil, T., Jones, N., Bird, K., Larson, A., Presler-Marshall, E., & Valters, C. (2015). Women's voice and leadership in decision-making: Assessing the evidence. Overseas Development Institute. https://gsdrc.org/document-library/womens-voice-and-leadership-in-decision-making-assessing-the-evidence/1

Eagly, A. H. (2005). Achieving relational authenticity in leadership: Does gender matter?. *The Leadership Quarterly*, 16(3), 459–474.

Eagly, A. H., & Carli, L. L. (2003). The female leadership advantage: An evaluation of the evidence. *The Leadership Quarterly*, 14(6), 807–834. https://doi.org/10.1016/j.leaqua.2003.09.004

Eagly, A. H., & Carli, L. L. (2007). Women and the labyrinth of leadership. *Harvard Business Review*, 85(9), 62–71.

Eagly, A. H., & Heilman, M. E. (2016). Gender and leadership: Introduction to the special issue. *The Leadership Quarterly*, 27(3), 349–353. https://doi.org/10.1016/j.leaqua.2016.04.002

Edmondson, A. (2019). *The Fearless Organization: Creating Psychological Safety in the Workplace for Learning, Innovation, and Growth.* Wiley.

Equal Pay Act of 1963, Pub.L. 88-38, 77 Stat. 56 (1963).

Ewens, M., & Townsend, R. R. (2020). Are early-stage investors biased against women? *Journal of Financial Economics*, 135(3), 653–677. https://doi.org/10.1016/j.jfineco.2019.11.004

Fairlie, R. W., & Robb, A. M. (2007). Why are Black owned businesses less successful than White owned businesses? The role of families, inheritances, and business human capital. *Journal of Labor Economics*, 25(2), 289–323.

Fisher, G., Graham, M., & Compeau, M. (2015). Women in leadership: The innovation perspective. In R. J. Burke & D. A. Major (Eds.), *Gender in Organizations: Are Men Allies or Adversaries to Women's Career Advancement?* (pp. 342–362). Edward Elgar Publishing.

Flex Ltd. (2023). Sustainability Report 2023: Net zero greenhouse gas emissions commitment & operational carbon reduction goals. Flex Ltd. https://flex.com/company/sustainability/net-zero-greenhouse-gas-emissions

Flexner, E., & Fitzpatrick, E. (1996). *Century of Struggle: The Woman's Rights Movement in the United States.* Harvard University Press.

Forbes. (2012, August 22). The world's most powerful women in business 2012: Ursula Burns. Forbes.

Gartner. (2021). 9 future of work trends post-COVID-19. Gartner Research. https://www.gartner.com/en/articles/9-future-of-work-trends-post-covid-19

Georgetown University. (2021). Stephanie Murphy (MSFS '04) becomes first Vietnamese American woman elected to Congress. https://msfs.georgetown.edu/news/stephanie-murphy/

Girls Who Code. (2023). Annual impact report: Closing the gender gap in tech. https://girlswhocode.com/media-kit/

Goleman, D. (2000). Leadership that gets results. *Harvard Business Review*, 78(2), 78–90. https://hbr.org/2000/03/leadership-that-gets-results

Gonzalez, A., & Clark, M. (2022). The impact of mentorship on women entrepreneurs in the U.S. innovation economy. U.S. Small Business Administration. https://advocacy.sba.gov/research/the-impact-of-mentorship-women-innovation/

Hancock, A.-M. (2016). *Intersectionality: An Intellectual History.* Oxford University Press.

Hill, E. J., Hawkins, A. J., Ferris, M., & Weitzman, M. (2001). Finding an extra day a week: The positive influence of perceived job flexibility on work and family life balance. *Family Relations*, 50(1), 49–58. https://doi.org/10.1111/j.1741-3729.2001.00049.x

Hossain, M. (2016). Strategic leadership and innovation: A case study of Ursula Burns at Xerox. *International Journal of Business and Management*, 11(6), 240–250.

Houser, K., Bowman Williams, J., & Submitter, G. L. C. (2021). Board gender diversity: A path to achieving substantive equality in the U.S. *SSRN Electronic Journal*, 63. https://doi.org/10.2139/ssrn.3796137

Hunt, V., Prince, S., Dixon-Fyle, S., & Dolan, K. (2020). *Diversity Wins: How Inclusion Matters.* McKinsey & Company. https://www.mckinsey.com/featured-insights/diversity-and-inclusion/diversity-wins-how-inclusion-matters

Infrastructure Investment and Jobs Act (2021). https://www.congress.gov/bill/117th-congress/house-bill/3684

Institute for Women's Policy Research. (2024, September). Fact sheet: Gender and racial wage gaps worsened in 2023 and pay equity still decades away (IWPR #C527). https://iwpr.org/wp-content/uploads/2024/09/IWPR-National-Wage-Gap-Fact-Sheet-2024.pdf

International Labour Organization. (2020). World employment and social outlook: Trends 2020. ILO. https://www.ilo.org/global/research/global-reports/weso/2020/lang--en/index.htm

Jasmin, J. (2017). Diversity in leadership and its organizational outcomes: A structured review and synthesis. *Journal of Organizational Change Management*, 30(7), 1119–1134.

Kamberidou, I. (2020). "Distinguished" women entrepreneurs in the digital economy and the multitasking whirlpool. *Journal of Innovation and Entrepreneurship*, 9(1). https://doi.org/10.1186/s13731-020-0114-y

Kanze, D., Huang, L., Conley, M. A., & Higgins, E. T. (2018). We ask men to win and women not to lose: Closing the gender gap in startup funding. *Harvard Business Review*, 96(5), 66–73.

Kiron, D., Unruh, G., Kruschwitz, N., Reeves, M., Rubel, H., & Fuisz-Kehrbach, S.-K. (2017). Corporate sustainability at a crossroads. *MIT Sloan Management Review*, 58(2), 1–30. https://sloanreview.mit.edu/article/corporate-sustainability-at-a-crossroads/

Knowledge@Wharton. (2010). *Xerox's Ursula Burns: Growing Up in a Housing Project – and Running a Fortune 500 Company*. Knowledge@Wharton.

Koenig, A. M., Eagly, A. H., Mitchell, A. A., & Ristikari, T. (2011). Are leader stereotypes masculine? A meta-analysis of three research paradigms. *Psychological Bulletin*, 137(4), 616–642. https://doi.org/10.1037/a0023557

Kossek, E. E., Lautsch, B. A., & Eaton, S. C. (2006). Telecommuting, control, and boundary management: Correlates of policy use and practice, job control, and work-family effectiveness. *Journal of Vocational Behavior*, 68(2), 347–367. https://doi.org/10.1016/j.jvb.2005.07.002

Kuffman Foundation. (2023). Kauffman indicators of entrepreneurship: Trends in women's entrepreneurship. https://indicators.kauffman.org/

Kumari, S., & Nair, J. (2023, September 11). Lisa Su's strategic leadership: Will she change her approach post Xilinx and Pensando acquisitions? *The Case for Women*, 1–27. https://doi.org/10.1108/CFW-07-2022-0020

Latu, I. M., Schmid Mast, M., Lammers, J., & Bombari, D. (2013). Successful female leaders empower women's behavior in leadership tasks. *Journal of Experimental Social Psychology*, 49(3), 444–448. https://doi.org/10.1016/j.jesp.2013.01.003

Lawless, J. L., & Fox, R. L. (2020). *It Still Takes a Candidate: Why Women Don't Run for Office* (3rd ed.). Cambridge University Press. https://doi.org/10.1017/CBO9780511778797

Li, F.-F. (2024). Profile of Fei-Fei Li. Stanford HAI. https://hai.stanford.edu/people/fei-fei-li

Lucey, C., & Burke, J. (2022). *Positive leadership in practice: a model for our future*. Routledge.

Madsen, S. R., & Andrade, M. S. (2018). Unconscious gender bias: Implications for women's leadership development. *Journal of Leadership Studies*, 12(1), 62–67. https://doi.org/10.1002/jls.21566

Magnanelli, B. S., & Raoli, E. (2025). Female leadership. In: Agostini, M., Beretta, V., Demartini, M.C., Ghio, A., Trucco, S. (eds) *Diversity and Equity in Accounting. SIDREA Series in Accounting and Business Administration*. Springer, Cham. https://doi.org/10.1007/978-3-031-78247-3_12

Marques, C. S., & Ferreira, J. J. (2021). Women entrepreneurs and firm performance: A systematic review. *Journal of Business Research*, 134, 1–15. https://doi.org/10.1016/j.jbusres.2021.05.031

McClain, P. D., & Stewart, J. (2019). *Can We All Get Along? Racial and Ethnic Minorities in American Politics* (7th ed.). Routledge.

McGill University. (2021, August 24). Dr. Joanne Liu joins McGill University. School of Population and Global Health, McGill University. https://www.mcgill.ca/spgh/article/dr-joanne-liu-joins-mcgill-university 1

McKinsey & Company. (2020). Diversity wins: How inclusion matters. https://www.mckinsey.com/featured-insights/diversity-and-inclusion/diversity-wins-how-inclusion-matters

McKinsey & Company & LeanIn.Org. (2020). *Women in the Workplace 2020*. McKinsey & Company and LeanIn.Org. https://wiw-report.s3.amazonaws.com/Women_in_the_Workplace_2020.pdf

McKinsey & Company & LeanIn.Org. (2021). Women in the workplace report. https://womenintheworkplace.com

McKinsey Global Institute. (2025). Women in tech: Closing the leadership gap. McKinsey & Company. https://www.mckinsey.com/mgi/publications/women-in-tech-2025

Migration Policy Institute. (2022). Immigrant health-care workers in the United States. https://www.migrationpolicy.org/article/immigrant-health-care-workers-united-states-2021

Moss-Racusin, C. A., Dovidio, J. F., Brescoll, V. L., Graham, M. J., & Handelsman, J. (2012). Science faculty's subtle gender biases favor male students. *Proceedings of the National Academy of Sciences*, 109(41), 16474–16479. https://doi.org/10.1073/pnas.1211286109

NASSCOM. (2024). Diversity & inclusion in Indian tech sector 2024. National Association of Software and Service Companies. https://www.nasscom.in/insights/reports

National Archives. (n.d.). 19th amendment to the U.S. constitution: Women's right to vote (1920). U.S. National Archives. https://www.archives.gov/milestone-documents/19th-amendment

National CIO Review. (2025, February 19). The AI rebellion has a new leader and her name is Mira Murati. The National CIO Review. https://nationalcioreview.com/articles-insights/technology/artificial-intelligence/the-ai-rebellion-has-a-new-leader-and-her-name-is-mira-murati/

National Institutes of Health. (2023). NIH workforce demographics. https://www.edi.nih.gov/services/workforce-data-insights/NIH-total-workforce-demographics

National Science and Technology Council. (2023). National artificial intelligence research and development strategic Plan: 2023 Update (OSTP Report). https://www.nitrd.gov/pubs/National-Artificial-Intelligence-Research-and-Development-Strategic-Plan-2023-Update.pdf 23

National Science Board. (2022). Science and Engineering Indicators 2022. https://ncses.nsf.gov/pubs/nsb20221/

National Science Foundation. (2007). *Women, Minorities, and Persons with Disabilities in Science and Engineering: 2007 (NSF 07-315)*. Arlington, VA. https://www.nsf.gov/statistics/wmpd/

National Science Foundation. (2023a). ADVANCE: Organizational change for gender equity in STEM academic professions. https://www.nsf.gov/funding/pgm_summ.jsp?pims_id=5383

National Science Foundation. (2023b). Mentorship and entrepreneurial persistence among women founders: Longitudinal data briefing. https://www.nsf.gov/statistics/mentorship_women_startups_2023.pdf

NCWIT. (2023). Scorecard 2023: Trends in women's retention in computing careers. https://ncwit.org/resources/scorecard-2023-women-retention/

Nguyen, G. T., Ly, H. T. M., Nguyen, T. H. H., & Hoang, S. (2024). Innovation leadership and workplace innovative behaviour: An experimental investigation of leadership dynamics among it engineers in Vietnam. *International Journal of Innovation Management*, *28*(05n06), 2450022.

Nishii, L. H. (2013). The benefits of climate for inclusion for gender-diverse groups. *Academy of Management Journal*, 56(6), 1754–1774. https://doi.org/10.5465/amj.2009.0823

Nobel Prize. (2023, October 2). The Nobel Prize in Physiology or Medicine 2023: Katalin Karikó – Facts. Nobel Prize Outreach AB. https://www.nobelprize.org/prizes/medicine/2023/kariko/facts/

Nooyi, I. (2021). *My Life in Full: Work, Family and Our Future*. Hachette UK.

NSF. (2023). Mentorship networks and entrepreneurial outcomes: Findings from NSF Innovation Corps (I-Corps). https://www.nsf.gov

OECD. (2024). The pursuit of gender equality: An uphill battle—2024 update. Organisation for Economic Co-operation and Development. https://www.oecd.org/gender

Organisation for Economic Co-operation and Development. (2022). *Gender equality and the empowerment of women and girls: Report*. OECD. https://www.oecd.org/content/dam/oecd/en/publications/reports/2022/05/gender-equality-and-the-empowerment-of-women-and-girls_9fbdf09e/0bddfa8f-en.pdf

OSTP Director Arati Prabhakar. (2023, March 3). Remarks of OSTP Director Arati Prabhakar at the 2023 AAAS Annual Meeting. White House Office of Science and Technology Policy. https://bidenwhitehouse.archives.gov/ostp/news-updates/2023/03/03/remarks-of-ostp-director-arati-prabhakar-at-the-2023-aaas-annual-meeting/

PCAST. (2024, October 15). Frances Colón, PhD | President's Council of Advisors on Science and Technology (PCAST). White House Office of Science and Technology Policy. https://bidenwhitehouse.archives.gov/pcast/members/frances-colon/ 56

PepsiCo. (2017). Annual reports. https://www.pepsico.com/investors/financial-information/annual-reports

Pew Research Center. (2015, January 14). *Women and Leadership*. Pew Research Center. https://www.pewresearch.org/social-trends/2015/01/14/women-and-leadership/

PitchBook. (2023). US VC female founders dashboard. https://pitchbook.com/news/articles/female-founders-venture-capital-2023

Population Reference Bureau. (2008, December). 2007 Occupational profiles reveal wide gender, racial gaps in science and engineering employment. https://www.prb.org/resources/2007-occupational-profiles-reveal-wide-gender-racial-gaps-in-science-and-engineering-employment/

Post, C., & Byron, K. (2015). Women on boards and firm financial performance: A meta-analysis. *Academy of Management Journal*, 58(5), 1546–1571. https://doi.org/10.5465/amj.2013.0319

Prabhakar, A. (2023). Let's change what's possible. *Science*, 379(6635), 1069. https://doi.org/10.1126/science.adh5328

Purdie-Vaughns, V., & Eibach, R. P. (2008). Intersectional invisibility: The distinctive advantages and disadvantages of multiple subordinate-group identities. *Sex Roles*, 59(5–6), 377–391.

Purkayastha, B. (2020). The unfinished journey from suffrage to substantive human rights: The continuing journey for racially marginalized women. *Western New England Law Review*, 42(3), 32.

PwC. (2024). *Women in Work 2024: International Mobility and Gender Diversity*. PricewaterhouseCoopers. https://www.pwc.com/womeninwork2024

Rhode, D. L. (2017). *Women and Leadership*. Oxford University Press.

Rholetter, W. (2024). *Ursula Burns*. In EBSCO Research Starters. EBSCO Information Services

Ryan, M.K., & Haslam, S.A. (2005). The glass cliff: Evidence that women are over-represented in precarious leadership positions. *British Journal of Management*, 16(2), 81–90.

Sanbonmatsu, K. (2019). Representation and recruitment: Women in U.S. politics. *Political Science Quarterly*, 134(3), 409–433. https://doi.org/10.1002/polq.12842

Sandretto, D., Liu, M., & Hernandez, L. (2025). Gender diversity leadership and ESG performance: The influence of women on boards and in management. *Business Strategy and the Environment*, 34(4), 5075–5094.

Santos, F. M., Pache, A.-C., & Birkholz, C. (2021). Making hybrids work: Aligning business models and organizational design for social enterprises. *California Management Review*, 63(1), 34–57. https://doi.org/10.1525/cmr.2015.57.3.36

Schwindt-Bayer, L. A. (2010). *Political Power and Women's Representation in Latin America*. Oxford University Press.

Smooth, W. G. (2021). Intersectionality and the election of Vice President Kamala Harris. *Politics & Gender*, 17(1), 163–167.

Spencer Stuart. (2023). *2023 S&P 500 New Director and Diversity Snapshot*. Spencer Stuart. https://www.spencerstuart.com/-/media/2023/august/usbi2023/director_diversity/2023_new_director_-diversity_snapshot.pdf 1

Stanford Humanities Center. (n.d.). *Joanne Liu*. Stanford Humanities Center. https://shc.stanford.edu/stanford-humanities-center/about/people/joanne-liu

Sue, D. W., Capodilupo, C. M., Torino, G. C., Bucceri, J. M., Holder, A. M. B., Nadal, K. L., & Esquilin, M. (2007). Racial microaggressions in everyday life: Implications for clinical practice. *American Psychologist*, 62(4), 271–286. https://doi.org/10.1037/0003-066X.62.4.271

Sweeney, L. (2013). Discrimination in online ad delivery. *Communications of the ACM*, 56(5), 44–54.

Techstars. (2022). Techstars Female Founders Fund. https://www.techstars.com/accelerators/female-founders-fund

Terjesen, S., Sealy, R., & Singh, V. (2009). Women directors on corporate boards: A review and research agenda. *Corporate Governance: An International Review*, 17(3), 320–337. https://doi.org/10.1111/j.1467-8683.2009.00742.x

Terjesen, S., Bosma, N., & Stam, E. (2015). Advancing public policy for high-growth, female, and social entrepreneurs. *Public Administration Review*, 76(2), 230–239. https://doi.org/10.1111/puar.12472

The White House. (2023, March 8). Vice President Kamala Harris celebrates progress on equity and women's rights. https://www.whitehouse.gov/briefing-room/statements-releases/2023/03/08/vice-president-kamala-harris-celebrates-progress-on-equity-and-womens-rights/

Thompson, A. C. (2022, July 12). Stephanie Murphy speaks on bipartisan cooperation during Jan. 6 hearings. ProPublica. https://www.propublica.org

TIME. (2023, March 5). This Fortune 500 CEO is obsessed with gender parity—here's what she's doing about it. TIME. https://time.com/6259808/revathi-advaithi-flex-ceo-gender-parity/

Tolan, S., et al. (2023). Inclusive program design and funding sustainability for women entrepreneurs: An intersectional approach. *Journal of Business Venturing Insights*, 20, e00326.

Trellis. (2024). Marilyn Waite: Managing director, Climate Finance Fund. https://trellis.net/article/author/marilyn-waite/

USAID (2023) https://www.usaid.gov/

U.S. Census Bureau. (2021, January 27). Women making gains in STEM occupations but still underrepresented. https://www.census.gov/library/stories/2021/01/women-making-gains-in-stem-occupations-but-still-underrepresented.html

U.S. Census Bureau. (2022). *What Is the Gender Wage Gap in Your State?* United States Department of Commerce. https://www.census.gov/library/stories/2022/03/what-is-the-gender-wage-gap-in-your-state.html

U.S. Citizenship and Immigration Services. (2024). *Immigration and Citizenship Data: Visa Processing, Green Card Backlog, and Policy Updates*. U.S. Department of Homeland Security. https://www.uscis.gov/tools/reports-and-studies/immigration-and-citizenship-data

U.S. Department of Energy. (2023). Inflation Reduction Act of 2022. https://www.energy.gov/lpo/inflation-reduction-act-2022

U.S. Department of Energy, Office of Economic Impact and Diversity. (2023). Diversity and inclusion in U.S. energy innovation. https://www.energy.gov/diversity/office-economic-impact-and-diversity

U.S. Department of Labor. (2023). Equal pay act of 1963 fact sheet. https://www.dol.gov/agencies/oasam/centennial/timeline/epa

U.S. House of Representatives. (n.d.). Congresswoman Stephanie Murphy. History, Art & Archives. https://history.house.gov/People/Detail/25769817292

UN Women. (2023, November 7). United Nations Executive Board report on gender equality. https://digitallibrary.un.org/nanna/record/4026844/files/UNW_2023_9-EN.pdf

UN Women Australia. (2023). Annual report 2023. https://unwomen.org.au/annual-reports-and-financial-statements/annual-report-2023/

United Nations Entity for Gender Equality and the Empowerment of Women (UN Women). (2023). Report on the second regular session 2023, 12 to 13 September 2023: Executive Board. UN Women. https://digitallibrary.un.org/nanna/record/4026844/files/UNW_2023_9-EN.pdf

Waite, M. (2021). *Sustainability at Work: Careers that Make a Difference*. Routledge.

Wang, F., & Karia, N. (2024). The impact of female leadership on innovation performance: A review study. *Global Business and Management Research: An International Journal*, 16(4s), 1564–1571. https://www.gbmrjournal.com/pdf/v16n4s/V16N4s-97.pdf

Wang, K. T., Cui, L., Zhu, N. Z., & Collier, P. (2024). Board gender diversity reforms around the world: The impact on corporate innovation. *Organization Science*. https://doi.org/10.1287/orsc.22.16956

West, J. (2020, July 3). Senior Director, International Climate Policy, Center for American Progress. Aspen Ideas. https://www.aspenideas.org/speakers/frances-colon

West, S. M., Whittaker, M., & Crawford, K. (2019). Discriminating systems: Gender, race, and power in AI. AI Now Institute Annual Report.

White House (2023). https://bidenwhitehouse.archives.gov/briefing-room/statements-releases/2023/06/22/joint-statement-from-the-united-states-and-india/

Woetzel, J., Madgavkar, A., Ellingrud, K., Labaye, E., Devillard, S., Kutcher, E., Manyika, J., Dobbs, R., & Krishnan, M. (2015). The power of parity: How advancing women's equality can add $12 trillion to global growth. McKinsey Global Institute. https://www.mckinsey.com/featured-insights/employment-and-growth/how-advancing-womens-equality-can-add-12-trillion-to-global-growth

Wolbrecht, C., & Campbell, D. E. (2017). *A Century of Votes for Women: American Elections Since Suffrage*. Cambridge University Press

Women Who Tech. (2022). Women who tech challenge. https://womenwhotech.com/challenge

WomenLift Health. (n.d.). Joanne Liu. WomenLift Health. https://www.womenlifthealth.org/profile/joanne-liu/

World Economic Forum. (2023). Global gender gap REPORT 2023. World Economic Forum. https://www.weforum.org/reports/global-gender-gap-report-2023

World Economic Forum. (2024). Global gender gap report 2024. World Economic Forum. https://www.weforum.org/reports/global-gender-gap-report-2024/

World Economic Forum. (2025, June 3). Women founders and venture capital – some 2023 snapshots. https://www.weforum.org/stories/2024/03/women-startups-vc-funding/

World Health Organization. (2023, December 13). WHO's Science Council issues report on mRNA vaccine technology. World Health Organization. https://www.who.int/news/item/13-12-2023-who-s-science-council-issues-report-on-mrna-vaccine-technology

Xu, Y. J. (2008). Gender disparity in STEM disciplines: A study of faculty attrition and turnover intentions. *Research in Higher Education*, 49(7), 607–624.

Zenger, J., & Folkman, J. (2020, April 15). Women score higher than men in most leadership skills. *Harvard Business Review*. https://hbr.org/2020/01/research-women-score-higher-than-men-in-most-leadership-skills

Zhang, B., Dafoe, A., & Dafoe, L. (2021). Diversity in artificial intelligence research. AI Index Report 2021, Stanford University.

8 Breaking Barriers
Women's Impact on Technology and Innovation Ecosystem

Rashmi Singh, Anuja Shukla and Sushil Kumar

8.1 INTRODUCTION

The technology and innovation ecosystem has traditionally been male dominated, with structural barriers and cultural norms limiting the participation of women across the globe. However, recent years have witnessed a transformative shift. Women are not only entering the tech space in increasing numbers but are also emerging as influential innovators, entrepreneurs, and leaders. Their increasing numbers are transforming the table of innovation through varied vision, previously untapped market demand, and inclusive economic development.

Various studies point out that women entrepreneurs, particularly in developing economies, are managing intricate intersections of gender, culture, and institutional issues (Onoshakpor et al., 2025; Shah et al., 2025). In spite of limitations, such entrepreneurs are capitalizing on digital platforms, social media, and new business models to mitigate resource shortages and social pressures (Al-Boinin et al., 2025). Programs on financial inclusion, capacity building, and policy support are also enabling women to scale their businesses and contribute to technology-led innovation (Medina-Vidal et al., 2025; Sahu et al., 2025).

Nevertheless, systematic biases persist to affect the access to capital, mentorship, and exposure for women in tech. Evidence indicates the existence of gender double standards in startup funding spaces, with women subjected to stringent tests and limited access to venture capital (Zhang & Chen, 2025). Foreign women especially have their challenges multiplied through language, culture, and institutional obstacles (Vučković et al., 2025).

This chapter aims to examine the role of women in the technology and innovation economy, how they are shattering conventional barriers, what challenges they uniquely encounter, and what are the factors that facilitate their progress. From drawing on contemporary literature and case studies, we hope to offer a thorough analysis of how not only are women entrepreneurs engaging in but also reshaping the innovation economy. This research seeks to examine the emerging role of women in the technology and innovation landscape through a systematic review and

 DOI: 10.1201/9781003604402-8

examination of contemporary Scopus-indexed research articles. The precise objectives are:

1. To identify and examine the emerging paramount barriers noted in the literature that women experience in technology and innovation fields, encompassing institutional, cultural, and fiscal barriers.
2. To analyze repeated themes and strategies mentioned in studies on how women entrepreneurs and innovators address such challenges.
3. To evaluate the impact of women-led innovation on technology, entrepreneurship, and socio-economic outcomes, as reported in recent studies.
4. To map out enabling factors such as digital tools, policy frameworks, and support systems that have been found to facilitate women's participation and leadership in innovative ecosystems.

8.2 LITERATURE REVIEW

A study was conducted among the Qatari women which talks about the interconnection among technology, gender, and culture acceptance in shaping entrepreneurial activities (Al-Boinin et al., 2025). According to the TAM (technology acceptance model), the different social media platforms offer autonomy and visibility, which is deeply engrossed with cultural and patriarchal norms to restrict full entrepreneurial freedom. The different financial inclusion model was specifically designed to serve women and other vulnerable populations (Medina-Vidal et al., 2025). Table 8.1 represents the highlights of different research contributions on women entrepreneurship.

8.2.1 Historical Context and Current Status

The involvement of women in technology and innovation has evolved significantly over time. In the 19th century, women like Ada Lovelace made pioneering contributions to computing by conceptualizing the first algorithm intended for a machine (Margolis & Fisher, 2002). The mid-20th century saw women playing crucial roles in computing during World War II, exemplified by the work of the Electronic Numerical Integrator and Computer (ENIAC) programmers (Light, 1999).

In spite of impressive historical achievements by women, they are still underrepresented in the tech industry. Women made up only 25% of the computing workforce as of 2020, with much lower numbers in leadership positions (NCWIT, 2020). The underrepresentation cuts across software development, engineering, and IT management, indicating entrenched gender disparities.

8.2.2 Challenges

One of the most cited challenges for women entrepreneurs is **access to capital**. Studies indicate that women receive less venture capital funding than men, often due to investor biases and risk perceptions (Eddleston, Ladge, Mitteness, & Balachandra, 2016). Women are also less likely to receive bank loans and more likely to self-finance their ventures, limiting their ability to scale operations (Robb & Coleman, 2010).

TABLE 8.1
Recent Research Contributions on Women Entrepreneurship[a]

No.	Title	Authors	Year	Journal	Key Findings (Summary)
1	Gender, Culture, and Social Media: Exploring Women Entrepreneurs in the Middle East	Al-Boinin H., Vatanasakdakul S., Zaghouani W.	2025	*Administrative Sciences*	Investigates how gender, culture, and TAM shape women's entrepreneurial use of social media in Qatar.
2	Financial inclusion of vulnerable sectors with a gender focus	Medina-Vidal A. et al.	2025	*Journal of Innovation and Entrepreneurship*	Proposes a financial inclusion model considering women and vulnerable groups.
3	Double standards in roadshows: Impact of investor gender bias	Zhang Q., Chen S.	2025	*International Journal of Entrepreneurial Behavior & Research*	Explores gender-based double standards during funding pitches.
4	Subsistence entrepreneurship and intersectionality	Shah U., Hayes N., Obaid A.	2025	*International Journal of Entrepreneurial Behavior & Research*	Examines how gender and class intersect in subsistence women entrepreneurs.
5	Challenges of migrant women entrepreneurs in Europe	Vučković V. et al.	2025	*Journal on Ethnopolitics and Minority Issues in Europe*	Identifies structural barriers migrant women face in entrepreneurship.
6	Entrepreneurship under patriarchy: Nigerian women's experiences	Onoshakpor C., Cunningham J., Gammie E.	2025	*International Journal of Entrepreneurial Behavior & Research*	Discusses constraints of patriarchy in women's entrepreneurial journeys in Nigeria.
7	"Look, Dear! You've Made a Discovery": Gender in Brazilian science history	Pires L.D.N., Peduzzi L.O.Q.	2025	*Revista Brasileira de Historia da Ciencia*	Investigates the marginalization of women's contributions to science and entrepreneurship in Brazil.
8	Capacity building for women entrepreneurs in Pakistan	Sahu P. et al.	2025	*Sustainable Entrepreneurship and Family Business*	Highlights skill development programs and policy gaps.
9	Female entrepreneurship in Italy: Institutional drivers	Nevi G. et al.	2025	*International Entrepreneurship and Management Journal*	Analyzes institutional enablers and barriers for women entrepreneurs.
10	"There is no limit to what we as women can accomplish": Hospitality sector analysis	Khalid R. et al.	2025	*Journal of Hospitality and Tourism Insights*	Investigates gender inequality and empowerment strategies in hospitality entrepreneurship.

a Created by authors.

Another significant barrier is the lack of access to entrepreneurial networks and mentorship. These networks provide not only knowledge and moral support but also critical business opportunities. Research suggests that women often find themselves excluded from male-dominated entrepreneurial ecosystems, which hinders their ability to grow their businesses (McGowan, Redeker, Cooper, & Greenan, 2012).

Systemic gender discrimination remains a major obstacle. Organizational cultures tend to be masculine, and they manifest in masculine norms, which result in gendered hiring, performance reviews, and fewer opportunities for women to get promoted (Faulkner, 2009).

Mentorship and professional connections are essential for career development, but women are often denied these chances. Women in technology are less likely to have mentors than their male counterparts and are not provided with access to male-dominated networking circles, which may impede career advancement and lead to professional isolation (McGee, 2018; Wingfield, 2009).

The demanding nature of tech careers, combined with societal expectations around caregiving, poses work-life balance challenges for women. Inflexible workplace policies often force women to choose between career advancement and family responsibilities (Cech & Blair-Loy, 2014).

Despite these challenges, women have significantly contributed to technological innovation. Studies indicate that gender-mixed teams are more innovative and perform better, especially in coming up with inclusive and user-focused solutions (Hong & Page, 2004; Page, 2007). High-profile leaders like Sheryl Sandberg and Marissa Mayer are examples of the revolutionary effect of women in leadership positions, and organizations like Girls Who Code, founded by Reshma Saujani, are instrumental in inspiring young women to pursue careers in this field (Sax et al., 2017).

Attempts to bridge the tech gap between genders encompass educational, business, and policy-focused interventions:

> Educational Programs: Initiatives such as Girls Who Code assist in developing early aptitude and interest in technology among young girls, creating a pipeline of future professionals (Sax et al., 2017). Corporate Diversity Initiatives: Businesses are taking measures such as diversity training, inclusive workplace environments, and mentoring to cater to women in technology (Hewlett et al., 2013). Policy Actions: Government policies encouraging equal pay, legislations against discrimination, anti-discrimination legislation, and flexible work arrangements are directed to promote a fairer environment.
>
> (Cech & Blair-Loy, 2014)

In recent decades, women have increasingly pursued entrepreneurial ventures, contributing to economic growth, job creation, and INNOVATION. However, female entrepreneurs continue to face structural and societal barriers that hinder their success, including limited access to finance, underrepresentation in high-growth sectors, and gendered perceptions of leadership (Brush, de Bruin, & Welter, 2009).

8.2.3 The Role of Technology in Empowering Women Entrepreneurs

There has been a major shift in women-led enterprises in terms of technology which plays a transformative role in enabling and scaling women-led enterprises. Different digital media and digital platforms reduce the entry barriers by lowering the cost of starting a new business and increasing access to markets. For example, e-commerce platforms like Etsy and Shopify have enabled women to monetize crafts and services globally with minimal overhead (Martin, 2019). There is certain traditional gatekeeping structure in the male-dominated industries which has been eradicated by social media and the digital marketing which allows women entrepreneur to build brands and reach customers without using traditional marketing tools such as advertising channels (Fischer & Reuber, 2014). Moreover, mobile technology and fintech solutions offer alternative pathways to financial inclusion. Mobile banking and crowdfunding platforms provide access to capital and financial services that may otherwise be inaccessible through traditional banks (Demirgüç-Kunt et al., 2018). With the technological advancement, it opens a new avenue for the women who remain underrepresented in tech-based startups and high-growth innovation-driven enterprises. This gap is often attributed to limited participation in STEM fields and bias within the startup ecosystem (Alsos, Hytti, & Ljunggren, 2013). Women tech founders often encounter skepticism from investors and experience challenges in accessing technical talent and resources.

Despite these challenges, women-led tech startups have demonstrated high levels of innovation and profitability. Research shows that female founders tend to build more socially responsible and inclusive companies, with stronger employee engagement and customer focus (Kanze, Huang, Conley, & Higgins, 2018).

To bridge the gender gap in entrepreneurship, targeted policy interventions and ecosystem support are essential. Government programs that promote digital literacy, provide startup funding, and facilitate mentorship can significantly increase women's entrepreneurial participation. For example, initiatives like the Women's Entrepreneurship Development Programme by the ILO and SheTrades by ITC aim to connect women entrepreneurs with global markets using technology. Incubators and accelerators tailored to women, such as Women's Startup Lab and Google's Women Techmakers, provide supportive environments that address the unique challenges women face in tech entrepreneurship (Henry, Foss, & Ahl, 2016).

8.3 DATA COLLECTION

To gain a comprehensive understanding of scholarly work related to women's entrepreneurship, a structured literature search was conducted using the Scopus database, which is one of the largest abstract and citation databases of peer-reviewed literature across disciplines. The search query used was TITLE-ABS-KEY ("women entr*"). This Boolean query retrieves documents that contain the phrase "*women entr*"* in the title, abstract, or keywords. The asterisk (*) is a truncation symbol used to capture variations of the root word, such as entrepreneurship, entrepreneurs, entrepreneurial, etc. This inclusive approach ensures that the search captures all relevant studies that discuss women in the context of entrepreneurial activities. The query yielded 4,567 documents as of the latest search. The search results reflect a growing academic interest in the role of women in entrepreneurship, especially in recent years, where themes such as digital entrepreneurship, gender-based barriers, innovation ecosystems, social entrepreneurship, and policy interventions have gained prominence. From this broad set of 4,567 documents, a text mining approach was undertaken to provide rich insights into the evolving narratives around women entrepreneurs, particularly their strategies for overcoming structural barriers and their contributions to technological innovation.

8.4 RESULTS

8.4.1 Descriptive Results

8.4.1.1 Insights from the Data: Leading Authors Trends in Entrepreneurship Research (2004–2025)

The chart in Figure 8.1 reveals that Welter, F. and Brush, C.G. are leading with the highest number of documents, followed by Dana, L.P., Henry, C., and Ramadani, V., who are mid-level contributors. Further, Kaciak, E., Yousafzai, S., and Ahl, H., as well as McAdam, M., and Agarwal, S., have the lowest publication counts among the top ten.

8.4.1.2 Top Journals by Publication Volume in Entrepreneurship Research

The line graph in Figure 8.2 reveals the publication trends among the top five academic journals from 2004 to 2025. The most consistent, dominant, and notable publication activity is shown by the journal *International Journal of Gender and Entrepreneurship* with 177 publications, followed by *International Journal of Entrepreneurship and Small Business* (90 articles) and *International Journal of*

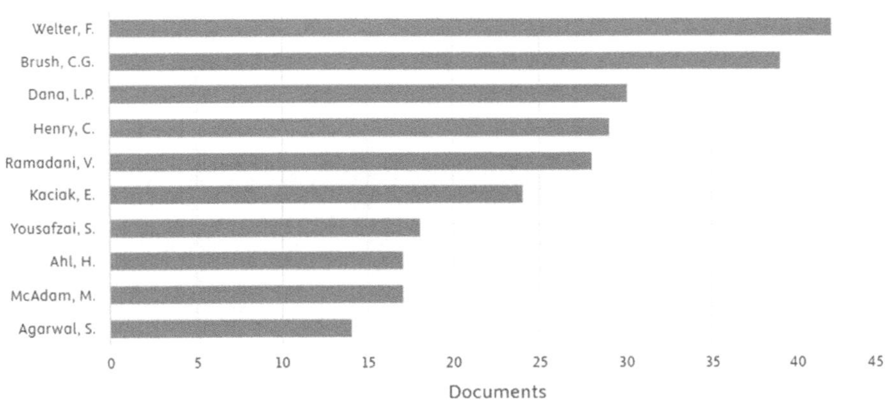

FIGURE 8.1 Leading authors' trends in entrepreneurship research (2004–2025).

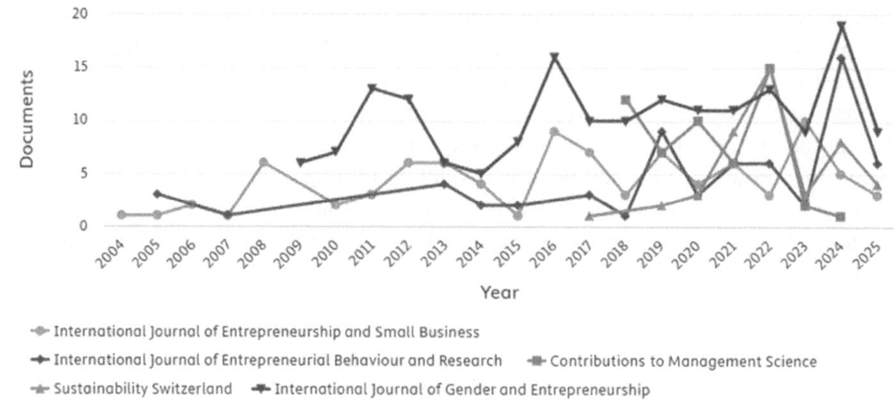

FIGURE 8.2 Leading journal trends in entrepreneurship research (2004–2025).

Entrepreneurial Behaviour and Research (64 articles). *Sustainability Switzerland* (45 articles) and *Contributions to Management Science* (53 articles) have shown significant growth post-2018, especially in 2022, indicating rising interest in sustainability and management-related entrepreneurship topics.

8.4.1.3 Top Academic Institutions Contributing to Women Entrepreneurship Research

Figure 8.3 represents the top academic institutions which contribute most to women entrepreneurship. Babson College is in a leading position which has significant 59 publications, indicating its strong academic focus and leadership in entrepreneurial studies. In the Middle Eastern and the Southeast European countries, **University of Tehran** and **Southeast European University** documented around 34 articles representing their active involvement. Furthermore, in the Germany and Sweden, Universität Siegen (Germany) and Jönköping University (Sweden) documented 32

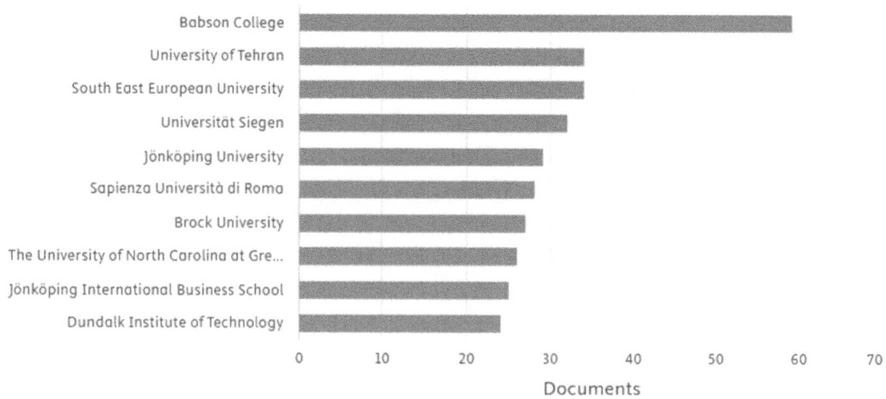

FIGURE 8.3 Leading universities in women entrepreneurship research output.

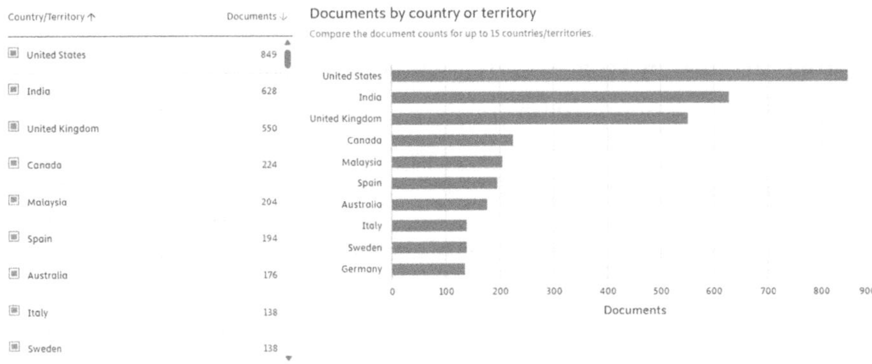

FIGURE 8.4 Global distribution of scholarly output on women entrepreneurship.

and 29 articles, respectively. This shows the diverse interests of researchers across the globe on a single research interest on women entrepreneurship. Babson College is the emerging dominant knowledge hub.

8.4.1.4 Top Contributing Countries in Women Entrepreneurship Research

As shown in Figure 8.4, top countries contributing to academic knowledge on women entrepreneurship are the USA, India, and the UK.

8.4.2 Text Mining Results

8.4.2.1 Topics

The authors have used different relevant topics which revolve around the women entrepreneurs, e.g., personal growth and life experiences, knowledge and learning, motivation and purpose, emotional and reflective storytelling. Different TED talks were selected because of this global connectivity and richness (Table 8.2). It is a credible source of insights with around 42.5M subscribers. This is a platform which features thought-provoking content from real-world practitioners, including women entrepreneurs, where they share their real-life journeys, experiences, failures, and lessons. These talks offer diverse perspectives on challenges, growth, and innovation, aligning well with the themes of personal development and entrepreneurial mindset. This is one of the most impactful mediums for understanding women's journeys in entrepreneurship.

8.4.2.2 Word Cloud

These are the top 10 Frequent Words: thing: 135, know: 127, time: 122, life: 113, one: 110, really: 95, right: 94, like: 93, day: 90, would: 78. The word cloud generator (Figure 8.5) and the frequency analysis highlight the most common used words in discussions related to women entrepreneurs. There are a few terms such as **"thing," "know," "time,"** and **"life"** that connect the personal narratives and reflective storytelling which are centered around emotion, experience, and growth. There are a few more words such as "one," "really," "right," "like," and "day" which indicate the strong motivational and emotional tone which reflects individual journeys,

TABLE 8.2
Inspiring TED Talks by Women Entrepreneurs and Leaders

S. No.	TED Talks	Links
1	Being an entrepreneur in a middle class family I Aditi Khattar I TEDxAIIMSRaipur	https://www.youtube.com/watch?v=SZ5ZDhCZNFc
2	How women entrepreneurs can drive growth of rural economy? I Miss. Manjari Sharma I TEDxIIMCalcutta	https://www.youtube.com/watch?v=QSACtsSbfwM
3	How to Transform Your Personality & Mindset I Bhavna Singh I TEDxSGTBKC	https://www.youtube.com/watch?v=NMfTtS7XT2w
4	Ideology of a woman entrepreneur I Minal Anand I TEDxYouth@GDGoenkaWorldSchool	https://www.youtube.com/watch?v=Jyad188iGvo
5	Challenging yourself is the key to achieving your endeavours I Spoorthi Vishwas I TEDxCITBengaluru	https://www.youtube.com/watch?v=VU7VIcd_i68
6	The Future of Woman Entrepreneurship I Shreya Bisht I TEDxMEC	https://www.youtube.com/watch?v=7BbGwradErU
7	The power of not thinking too much I Pranjal Priya I TEDxCCOEW	https://www.youtube.com/watch?v=SDqB_dhqWdI
8	Why consistency matters more than you think I Niharikaa Kaur Sodhi I TEDxDFBEDU	https://www.youtube.com/watch?v=LeIoaKLG2Js
9	Failure plus consistency equals success I Shikhi Sharma I TEDxIIS Youth	https://www.youtube.com/watch?v=WIv1TuuoeBw

FIGURE 8.5 Visual summary of commonly used words (author creation).

challenges, and pivotal moments. The overall word cloud talks about the aspirational language, decision-making, and hypothetical thinking. From the above insights, we conclude that women often articulate stories based on their personal experience, determination, and vision, emphasizing lived experience as a driving force in their entrepreneurial journeys.

8.4.2.3 Clusters

The text was clustered into four themes (Table 8.3).

Cluster 1: "Struggles & Dreams of Breaking Free"
Cluster 1 reflects emotional stories of escaping expectations, overcoming depression, and chasing larger dreams. The Top Terms are dream, going, job, country, big, happened, know, people, depression, India. Some sample reviews are explained below:

Sample 1: parents' life difficult by setting in my own set of rules looking back I don't think it's re…
Sample 2: for me immediately after my graduation I was offered an internship in a media house I was g…
Sample 3: and I logged myself up for two years I started feeling monotonous stagnant and I was fragile…

Cluster 2: "Uncertain Beginnings & Personal Growth"
Cluster 2 reflects uncertainty mixed with resilience. It includes Top Terms: time, much, year, able, wanted, action, faith, come, third, could. Some sample reviews are explained below:

Sample 1: upon me you know why because when a middle class science graduate shifts Kario to a non-sci…
Sample 2: page on Instagram and without knowing much and expecting much I started documenting whateve…
Sample 3: wanted it to be and you know there are days when you feel low and bad…

Cluster 3: "Identity, Resistance & Expression"
Cluster 3 implies resistance to norms and assertion of self. The Top Terms are: would, woman, day, like, know, thing, brand, really, right, one. Some sample reviews are explained below:

Sample 1: foreign first of all thank you so much to invite me on this platform I'm grateful I'll star…
Sample 2: without an iota of regret and this is my story I was born and raised in a typical Orthodox
Sample 3: all right when my parents and my grandfather got to know about my choice of Commerce there …

TABLE 8.3
Cluster

Cluster	Suggested Name	Reason
1	Struggles & Dreams of Breaking Free	Emotional struggle, dreams vs. societal expectations
2	Uncertain Beginnings & Personal Growth	Trial-and-error, early transitions, inner development
3	Identity, Resistance & Expression	Gender roles, defiance, personal voice, authenticity
4	Purpose & Navigating Expectations	Career-life balance, societal norms, reflective purpose

Cluster 4: "Purpose & Navigating Expectations"

Cluster 4 sits between introspection and pragmatism which talks about how to live a meaningful life while managing external pressure. The Top Terms are: life, work, thing, make, one, people, right, go, important, good. Some sample reviews are explained below:

Sample 1: if you get a good percentage you take science if you take a get an average percentage you t…

Sample 2: passions there were certain things I was passionate about and even though I could not pursu…

Sample 3: it enjoyed it adding to my skill set adding to my organization and I was very well supporte…

8.5 IMPLICATION

The study findings highlight the importance of personal experiences, emotional resilience, and motivational narratives that frames the journey of women entrepreneurs. There are different themes, e.g., knowledge, acquisition, and aspirational thinking, which talk about the individual personality and the different life experiences/roles which are deeply intertwined with their entrepreneurial success. This is an emerging topic which grabs attention among the scholars globally. There are different research and institutional contributions globally which talk about challenges and opportunities, experiences at different stages of women's entrepreneurial journey. Furthermore, different multimedia sources such as TED Talks enrich understanding by providing authentic, real-world perspectives that complement traditional academic findings. Together, these insights emphasize the importance of integrating personal narratives and socio-cultural factors into entrepreneurship research and support mechanisms.

8.5.1 Managerial Implications

Knowing the lived experiences of women entrepreneurs provides vital knowledge for organizations that seek to develop inclusive innovation ecosystems. The results highlight the multifaceted interplay of identity, resilience, autonomy, and purpose in women's entrepreneurial trajectories. With reference to these findings, the following managerial implications are put forward. Those who are holding managerial positions should be actively involved in nurturing the workplace culture in such a way that fosters inclusive and empowering work environments where women can express themselves and pursue entrepreneurial aspirations. There should be fair access to professional development opportunities such as developing mentorship schemes and inclusive leadership. Every organization should encourage diversity in voices which will enhance innovation and organizational flexibility. Resilience and mental well-being are the two components in which the company should focus on. Every organization should provide systematic support systems based on emotional and psychological needs of being an entrepreneur. Programs like resilience training, mental health programs, and stress management workshops are necessary to enable women to manage uncertainty and maintain well-being throughout their career. There should be flexible support and non-traditional career paths which will engage and hold women in innovation-oriented positions. In order to avoid work-family conflict, there should be different career path policies such as work from home, flexible working hours, and personalized career management, which allow women to balance individual development and family care with career progression. There should be a purpose-driven leadership, and the purpose should be aligned with organizations' mission and vision. It encourages motivation and path-oriented leadership, strengthens organization loyalty and develops a sense of belongingness which acts as a driver for long-term engagement and innovation. Enabling women entrepreneurship is not just an issue of equity but an organizational imperative for sustained innovation and growth. By creating inclusive workplaces, promoting well-being, facilitating flexibility, and instilling a sense of common purpose, managers can realize the full potential of women entrepreneurs and help generate real organizational change.

8.5.2 Academic Implications

This research offers valuable understanding of women entrepreneurs lived experiences and presents a number of avenues for future scholarly research. Tracing the psychological, social, and structural aspects of women's entrepreneurship, scholars can deepen the debate on gender, innovation, and enterprise development. First, academics need to give more importance to qualitative research that is able to capture women's entrepreneurial experiences in terms of their emotional and identity-related aspects. Those studies that address the way women negotiate societal expectations, articulate identities, and cope with emotional challenges—like anxiety, loneliness, and depression—can provide dense, context-specific information on their drivers and challenges. Second, there is a call for longitudinal and cross-disciplinary research that investigates the growth of resilience in women entrepreneurs. Based on disciplines like psychology, organizational behavior, and sociology, researchers can investigate how women construct adaptive capacities in risky contexts and how individual development interacts with entrepreneurial achievement. Third, scholarly research ought to explore in detail how women entrepreneurs reconcile personal purpose and professional life, especially in terms of balancing societal and cultural expectations. Purpose-driven entrepreneurship needs more conceptual scrutiny, especially against diverse sociocultural backgrounds and with diverse institutional constraints. Fourth, work-life balance continues to be a recurring challenge for most women entrepreneurs. Future studies should investigate methods that facilitate sustainable work-life integration, taking into account the cultural, economic, and industry contexts. Further studies should examine how institutional and organizational designs can be optimized to facilitate women's long-term engagement in innovation systems. The changing women's entrepreneurship environment requires increased scholarly focus that transcends economic impacts to encompass emotional, social, and cultural aspects. Through adopting integrative, multidisciplinary, and inclusive methods, research in the future can help develop more balanced and innovative entrepreneurial environments.

8.6 LIMITATION AND FUTURE RECOMMENDATION

This study revolves around the insights from valuable thematic analysis, text mining of existing literature, sentiment analysis, and public available talks, which may omit less documented or informal entrepreneurial experiences. There is one gap identified which suggests further qualitative research which could deepen the understanding of women entrepreneurs' emotional and psychological mindset. In future, researchers can focus on these two factors. Furthermore, researchers can incorporate longitudinal studies to talk about emotions and motivations that change over a period. A comparative cross-cultural study can also be conducted which will capture a diverse range of emotions and motivations. More constructive interviews, case studies, and different policy evaluations can provide a more comprehensive picture and inform targeted support programs for women entrepreneurs worldwide. While significant progress has been made, ongoing research is needed to understand the complex dynamics of gender in technology and innovation. Future studies should explore the intersectionality of gender with other identity factors, such as race and socioeconomic status, to gain a comprehensive understanding of the barriers women face (Crenshaw, 1989). Additionally, longitudinal studies could provide insights into the long-term effects of diversity initiatives on the innovation ecosystem.

REFERENCES

Al-Boinin, H., Vatanasakdakul, S., & Zaghouani, W. (2025). Gender, culture, and social media: Exploring women entrepreneurs in the Middle East. *Administrative Sciences*, 15(3), 89. https://doi.org/10.3390/admsci15030089

Alsos, G. A., Hytti, U., & Ljunggren, E. (2013). Gender and innovation: State of the art and a research agenda. *International Journal of Gender and Entrepreneurship*, 5(3), 236–256. https://doi.org/10.1108/IJGE-06-2013-0049

Brush, C. G., de Bruin, A., & Welter, F. (2009). A gender-aware framework for women's entrepreneurship. *International Journal of Gender and Entrepreneurship*, 1(1), 8–24. https://doi.org/10.1108/17566260910942318

Cech, E. A., & Blair-Loy, M. (2014). Consequences of flexibility stigma among academic scientists and engineers. *Work and Occupations*, 41(1), 86–110.

Crenshaw, K. (1989). Demarginalizing the intersection of race and sex: A black feminist critique of antidiscrimination doctrine, feminist theory, and antiracist politics. *University of Chicago Legal Forum*, 1989(1), 139–167.

Demirgüç-Kunt, A., Klapper, L., Singer, D., Ansar, S., & Hess, J. (2018). *The Global Findex Database 2017: Measuring Financial Inclusion and the Fintech Revolution*. World Bank.

Eddleston, K. A., Ladge, J. J., Mitteness, C., & Balachandra, L. (2016). Do you see what I see? Signaling effects of gender and firm characteristics on financing entrepreneurial ventures. *Entrepreneurship Theory and Practice*, 40(3), 489–514. https://doi.org/10.1111/etap.12117

Faulkner, W. (2009). Doing gender in engineering workplace cultures: Observations from the field. *Engineering Studies*, 1(1), 3–18.

Fischer, E., & Reuber, A. R. (2014). Online entrepreneurial communication: Mitigating uncertainty and increasing differentiation via Twitter. *Journal of Business Venturing*, 29(4), 565–583. https://doi.org/10.1016/j.jbusvent.2014.02.004

Henry, C., Foss, L., & Ahl, H. (2016). Gender and entrepreneurship research: A review of methodological approaches. *International Small Business Journal*, 34(3), 217–241. https://doi.org/10.1177/0266242614549779

Hewlett, S. A., Marshall, M., & Sherbin, L. (2013). *Harnessing the Power of the Purse: Female Investors and Global Opportunities for Growth*. Center for Talent Innovation.

Hong, L., & Page, S. E. (2004). Groups of diverse problem solvers can outperform groups of high-ability problem solvers. *Proceedings of the National Academy of Sciences*, 101(46), 16385–16389.

Kanze, D., Huang, L., Conley, M. A., & Higgins, E. T. (2018). We ask men to win and women not to lose: Closing the gender gap in startup funding. *Academy of Management Journal*, 61(2), 586–614. https://doi.org/10.5465/amj.2016.1215

Khalid, R., Alharthi, F., & Watson, K. (2025). "There is no limit to what we as women can accomplish": Hospitality sector analysis. *Journal of Hospitality and Tourism Insights*, 8(1), 47–62. https://doi.org/10.1108/JHTI-12-2023-0959

Light, J. S. (1999). When computers were women. *Technology and Culture*, 40(3), 455–483.

Margolis, J., & Fisher, A. (2002). *Unlocking the Clubhouse: Women in Computing*. MIT Press.

Martin, L. (2019). E-commerce and female entrepreneurs: Opportunities and challenges in a digital economy. *Journal of Business Research*, 100, 531–536. https://doi.org/10.1016/j.jbusres.2018.10.051

McGee, E. O. (2018). Devalued black and latino racial identities: A byproduct of STEM college culture? *American Educational Research Journal*, 53(6), 1626–1662.

McGowan, P., Redeker, C. L., Cooper, S. Y., & Greenan, K. (2012). Female entrepreneurship and the management of business and domestic roles: Motivations, expectations and realities. *Entrepreneurship & Regional Development*, 24(1–2), 53–72. https://doi.org/10.1080/08985626.2012.637351

Medina-Vidal, A., Prado-Román, C., & Muñoz, D. (2025). Financial inclusion of vulnerable sectors with a gender focus. *Journal of Innovation and Entrepreneurship*, 14, 63. https://doi.org/10.1186/s13731-025-00463-2

NCWIT (National Center for Women & Information Technology). (2020). By the numbers. Retrieved from https://ncwit.org/resources/numbers

Nevi, G., Testarmata, S., & Simoni, M. (2025). Female entrepreneurship in Italy: Institutional drivers and constraints. *International Entrepreneurship and Management Journal*, 21(2), 234–251. https://doi.org/10.1007/s11365-024-01012-1

Onoshakpor, C., Cunningham, J., & Gammie, E. (2025). Entrepreneurship under patriarchy: Nigerian women's experiences. *International Journal of Entrepreneurial Behavior & Research*, 31(1), 78–95. https://doi.org/10.1108/IJEBR-04-2023-0334

Page, S. E. (2007). *The Difference: How the Power of Diversity Creates Better Groups, Firms, Schools, and Societies*. Princeton University Press.

Pires, L. D. N., & Peduzzi, L. O. Q. (2025). "Look, Dear! You've made a discovery": Gender in Brazilian science history. *Revista Brasileira de História da Ciência*, 18(2), 102–119. https://doi.org/10.53727/rbhc.v18i1.1037

Robb, A. M., & Coleman, S. (2010). Financing strategies of new technology-based firms: A comparison of women- and men-owned firms. *Journal of Technology Management & Innovation*, 5(1), 30–50. https://doi.org/10.4067/S0718-27242010000100003

Sahu, P., Khan, M., & Iqbal, J. (2025). Capacity building for women entrepreneurs in Pakistan. In R. Ratten (Ed.), *Sustainable Entrepreneurship and Family Business* (pp. 183–202). IGI Global. DOI: 10.4018/979-8-3693-8694-1.ch011

Sax, L. J., Kanny, M. A., Riggers-Piehl, T. A., Whang, H., & Paulson, L. N. (2017). "But I'm not good at math": The changing salience of mathematical self-concept in shaping women's and men's STEM aspirations. *Research in Higher Education*, 56(8), 813–842.

Shah, U., Hayes, N., & Obaid, A. (2025). Subsistence entrepreneurship and intersectionality: The lived experiences of Pakistani women entrepreneurs. *International Journal of Entrepreneurial Behavior & Research*, 31(3), 145–162. https://doi.org/10.1108/IJEBR-12-2022-1094

Vučković, V., Potz, P., Woldemichael, Y. A., & Haydn, F. (2025). Challenges of migrant women entrepreneurs in Europe: Addressing structural barriers. *Journal on Ethnopolitics and Minority Issues in Europe*, 24(1), 56–74. https://doi.org/10.53779/VVRB1903

Wingfield, A. H. (2009). Racializing the glass escalator: Reconsidering men's experiences with women's work. *Gender & Society*, 23(1), 5–26.

Zhang, Q., & Chen, S. (2025). Double standards in roadshows: Impact of investor gender bias. *International Journal of Entrepreneurial Behavior & Research*, 31(2), 123–140. https://doi.org/10.1108/IJEBR-11-2023-1188

9 Women Shaping the Landscape of AI

Breaking Barriers

Jyoti Suri, Praveen Kumar Pandey and Mohd Asif Shah

9.1 INTRODUCTION

The path of women into leadership roles has been shaped by centuries of struggle, resilience, and transformation. For a long time, leadership—whether in politics, business, education, or religion—was seen as a man's prerogative. Women were expected to remain in the private sphere, managing homes and families, while decision-making and authority in the public realm were reserved almost exclusively for men. This divide wasn't accidental; it was built on deeply rooted patriarchal beliefs that equated leadership qualities like assertiveness, rationality, and authority with masculinity (Eagly & Carli, 2007).

Historically, systemic barriers—both visible and invisible—kept women away from leadership pathways. For instance, in many parts of the world, women were denied access to formal education, the right to vote, and the legal standing to own property or enter into contracts. These limitations deprived them of the essential tools needed for civic participation, economic independence, and upward mobility (Scott, 1996). Even when women worked outside the home, their roles were typically low-paid, low-status, and rarely came with any authority or decision-making power.

The first major cracks in this system appeared during the 19th and early 20th centuries with the rise of the women's suffrage movement. Women across the globe began to demand the right to vote, asserting that citizenship must include a voice in how society is governed. Leaders like Susan B. Anthony and Elizabeth Cady Stanton in the United States, and Emmeline Pankhurst in the United Kingdom, spearheaded movements that were often met with resistance, arrests, and ridicule. Yet their determination eventually bore fruit—the United States granted women the right to vote in 1920 through the 19th Amendment, and the UK followed suit in stages between 1918 and 1928 (Flexner & Fitzpatrick, 1996; Purvis, 1995).

However, gaining the vote was just the beginning. The real battle was about changing mindsets and structures that had long excluded women from leadership. The mid-20th century brought a new wave of feminism—often called the "second wave"—which expanded the conversation beyond legal rights. Women began to question why they were underrepresented in boardrooms, parliaments, and universities, and why domestic

DOI: 10.1201/9781003604402-9

and caregiving work was undervalued. The movement called for equal opportunities in education and employment, reproductive rights, and an end to gender-based discrimination at all levels of society (Freeman, 1975; Friedan, 1963).

This era saw the enactment of landmark legal reforms. In the United States, the Equal Pay Act of 1963 mandated equal wages for equal work, and Title IX of the Education Amendments of 1972 prohibited sex-based discrimination in any federally funded education program, opening up new doors for women in academic and athletic leadership (Baker, 1981). Similar efforts took place in other countries, including the Equal Pay Act 1970 in the UK and the implementation of gender quotas in Scandinavian parliaments. These policies began to address structural inequalities and created more visible pathways for women to rise as leaders.

Despite these advances, the journey has not been smooth. Even today, women in leadership often face what researchers term the "glass ceiling" or "labyrinth"—a complex set of barriers including gender stereotypes, work-life balance pressures, and exclusion from informal networks of power (Eagly & Carli, 2007). Yet, progress is undeniable. Women as the educational leaders, the social activists, the CEOs, the heads of state, are persisting to redefine leadership through providing various perspectives, collaborative styles, and inclusive practices. Essentially, the history of leadership among women is not a singular account of success, it is a shared vision of what a leader can be and how they must lead in the contemporary society.

9.2 SHIFTING PARADIGMS: WOMEN IN MODERN LEADERSHIP

Over the past few years, the dynamics of leadership have begun to alter. An increasing number of women are taking up powerful positions in business, politics, academia, and civil society like never before. While men still dominate the top tiers of leadership globally, the presence of women is steadily increasing. According to the World Economic Forum's Global Gender Gap Report (2023), women now hold about 32.2% of senior managerial roles worldwide. This marks progress, but true equality is still a long way off. In the political arena, women occupy only 26.5% of parliamentary seats globally—a number that reflects how male-dominated the world of politics continues to be (Inter-Parliamentary Union, 2023).

Yet, the impact of women who do reach leadership positions has been powerful. Research shows that women often lead differently—not because of their gender per se, but because of the social experiences and challenges they've had to navigate. Women leaders are more likely to adopt collaborative, empathetic, and inclusive approaches, which have been shown to enhance team performance, employee morale, and even innovation (Eagly, Johannesen-Schmidt, & van Engen, 2003).

The value of these leadership styles became especially clear during global crises. During the COVID-19 pandemic, for instance, women leaders like Jacinda Ardern in New Zealand and Angela Merkel in Germany gained international respect for their calm, science-based, and compassionate responses. A comparative study by Garikipati and Kambhampati (2020) found that countries led by women had, on average, better outcomes in controlling the spread of COVID-19, thanks in part to timely lockdowns and clear communication.

Moreover, women leaders bring something critical to decision-making: diverse lived experiences. Their perspectives help ensure that policies and strategies consider a broader range of social, economic, and cultural realities. In business, this kind of diversity isn't just about fairness—it also makes sense financially. A 2020 report by McKinsey & Company found that companies with higher gender diversity in leadership roles are more likely to outperform their less diverse counterparts in profitability and value creation.

Still, many challenges remain. The so-called "glass ceiling"—an invisible barrier that keeps women from reaching the highest levels—continues to exist in many sectors, especially in STEM fields, top corporate roles, and high-level politics, particularly in developing nations. Even when women do make it to the top, they often find themselves in risky or unstable positions—a phenomenon known as the "glass cliff." This means women are more likely to be given leadership roles during crises, where the chance of failure is higher (Ryan & Haslam, 2005). In essence, they're often set up to fail, or at least to struggle more than their male counterparts.

In addition, gender stereotypes and implicit biases continue to shape perceptions. For example, assertiveness in male leaders is often seen as strength, while in female leaders it might be labeled as aggression. At the same time, expectations around caregiving and work-life balance still fall disproportionately on women, making the climb to leadership even steeper.

Another critical factor that shapes women's leadership experiences is intersectionality—a concept introduced by legal scholar Kimberlé Crenshaw in 1991. Not all women face the same barriers. Women of color, women from marginalized castes or communities, LGBTQ+ women, and women with disabilities often encounter multiple layers of discrimination. Their journey to leadership is not only about gender but also about race, class, sexuality, and ability. Ignoring these overlapping challenges risks creating a leadership narrative that only benefits a privileged few.

Despite these hurdles, women are continuing to break boundaries. From heading Fortune 500 companies to leading grassroots movements, their leadership is redefining power—not as control, but as collaboration; not as dominance, but as transformation.

9.3 REDEFINING LEADERSHIP: TOWARD AN INCLUSIVE FUTURE

As more women ascend into leadership, they are not merely fitting into existing molds—they are transforming leadership norms. The emergence of new leadership models—such as servant leadership, transformational leadership, and authentic leadership—aligns with qualities often attributed to women leaders: empathy, collaboration, vision, and resilience (Northouse, 2021).

Mentorship programs, gender quotas, inclusive hiring policies, and flexible workplace practices are helping close the gender leadership gap. However, systemic transformation requires continued efforts at all levels—educational, organizational, and governmental. Only through sustained structural change can gender equity in leadership be fully realized.

9.3.1 Breaking Barriers Historically

The evolution of women in leadership has not been a linear journey—it has been a story marked by resilience, persistence, and a constant battle against systemic obstacles. Across centuries and cultures, women have challenged rigid gender roles, navigated hostile environments, and redefined what it means to lead. Despite widespread societal limitations, they have continuously made space for themselves at decision-making tables (Lerner, 1993; Scott, 1996).

9.4 PROMPT WOMEN LEADERS—CHALLENGES

1. **Cultural Barriers and Societal Norms**
   Throughout history, leadership has largely been viewed as a male preserve. Traits commonly associated with leadership—such as decisiveness, authority, and assertiveness—were long culturally coded as masculine (Eagly & Karau, 2002). Women who exhibited such traits were often criticized for being unfeminine or overly ambitious. They were pressured to conform to traditional domestic roles as wives, mothers, and caretakers, and those who deviated were frequently stigmatized. In many societies, particularly in the Global South, cultural expectations regarding women's "place" continue to shape leadership pipelines (Nussbaum, 2000; Ridgeway, 2001).
2. **Institutional and Legal Barriers**
   Formal institutions also played a role in keeping women out of leadership. Until the 20th century, many legal frameworks denied women the right to vote, own property, or even access education. In colonial and postcolonial India, for example, women were often subject to both patriarchal traditions and colonial legal systems that excluded them from civic life (Forbes, 1996). In the Western world, figures like Susan B. Anthony and Elizabeth Cady Stanton fought bitter battles for women's suffrage in the United States, leading to the 19th Amendment in 1920 (Flexner & Fitzpatrick, 1996). Similarly, in the UK, suffragettes led by Emmeline Pankhurst agitated for years before limited voting rights were granted to women in 1918 and equal suffrage in 1928 (Purvis, 1995). Without legal and institutional access, leadership was effectively out of reach.
3. **Lack of Networks and Mentorship**
   Another serious challenge for early women leaders was the absence of mentorship and professional networks. Leadership development typically occurs through informal channels—social capital, insider networks, and mentorship relationships. In male-dominated domains, women found themselves without sponsors or role models who could advocate for them (Ibarra, Carter, & Silva, 2010). Historical records indicate that many pioneering women had to build their credibility alone, often facing isolation or exclusion in their fields. The absence of visible female leaders created a vacuum of aspirational models for younger women, further perpetuating underrepresentation (Kanter, 1977).

4. **Equilibrium Maintenance—The Double Burden**
   Women have also faced what is commonly referred to as the "double burden"—the simultaneous pressure of balancing professional responsibilities with traditional domestic roles. Even as they pursued leadership, women were expected to continue fulfilling their roles as caregivers, homemakers, and emotional anchors of the family (Hochschild & Machung, 1989). This societal expectation limited their mobility, reduced available time for leadership engagement, and frequently forced women to make difficult trade-offs between career advancement and family obligations. The perception that women are less "available" for leadership-intensive roles still contributes to their underrepresentation in top-tier positions (Bianchi et al., 2000; Hewlett, 2002).

9.5 WOMEN AND ARTIFICIAL INTELLIGENCE: BREAKING NEW GROUND

In the 21st century, artificial intelligence (AI) stands at the forefront of global innovation. AI refers to the development of systems capable of performing tasks that typically require human intelligence—such as reasoning, decision-making, language processing, and pattern recognition (Russell & Norvig, 2020). From healthcare to finance, from education to governance, AI is reshaping how society functions. However, this technological revolution is not gender-neutral. Women remain significantly underrepresented in the AI ecosystem—across research, policy, corporate leadership, and technical roles (UNESCO, 2020).

Despite the dominance of men in modern AI, it is important to remember that women were foundational in the development of computing and early AI. Ada Lovelace, often cited as the first computer programmer, imagined the potential of computational machines in the 19th century (Toole, 1998). Grace Hopper developed the first compiler and was instrumental in the development of COBOL, a foundational programming language (Beyer, 2009). Yet, these contributions were often marginalized or excluded from mainstream tech histories.

Today, according to the *World Economic Forum's Global Gender Gap Report* (2023), women make up only 22% of AI professionals worldwide. This gender disparity stems from persistent issues—gender stereotypes in education, lack of encouragement to pursue STEM, and male-dominated academic and professional spaces (UNESCO, 2020; West, Kraut, & Ei Chew, 2019). These barriers reinforce the misconception that AI is a "male field," creating a self-perpetuating cycle of exclusion.

9.6 WHY REPRESENTATION IN AI MATTERS

AI is not created in a vacuum. The data used to train algorithms, the assumptions built into systems, and the ethical implications of deployment all reflect the biases of the teams designing them (O'Neil, 2016; Noble, 2018). If women—and especially women from diverse backgrounds—are not meaningfully included, the risk of embedding gender and racial biases into AI systems becomes dangerously high.

Biased AI applications in hiring, credit scoring, policing, and facial recognition have already had discriminatory outcomes (Buolamwini & Gebru, 2018).

Thus, increasing women's representation in AI is not just about fairness—it is critical to **technological accuracy, ethics, and innovation**. Diverse teams tend to develop more creative and equitable solutions. A McKinsey & Company report (2020) confirmed that companies with greater gender diversity in leadership were 25% more likely to achieve above-average profitability.

9.7 TOWARD INCLUSIVE AI LEADERSHIP

Building a more inclusive AI future requires targeted and systemic action:

1. **Education Solutions**: Girls have to be introduced to education programs, learn to break down gender stereotypes, in the field of math and science education, as well as being introduced to the career in the field of AI and data science, at a young age.
2. **Mentorship and Professional Networks**: The organizations should take up mentorship and professional network procedures, which help women develop academically and professionally (Hewlett et al., 2010).
3. **Policy Reform and Inclusion in Workplace**: Organizations must adopt inclusive recruitments, transparent promotion processes, generous maternity, and flexi-time arrangements to ensure that women stay in the industry (Berg et al., 2019).
4. **Honoring Women Pioneers**: By rewriting history, to commemorate the accomplishments of women in technology and AI, a potential future generation may become inspired to continue with the effort, and the historical oversights can be addressed (Abbate, 2012).
5. **Intersectional Treatment**: To make sure that no one is left behind in terms of discrimination, policies are needed that would focus on the intersectional discrimination that women of color, LGBTQ+ women, women living in rural areas, and women with disabilities have to endure (Crenshaw, 1991; Thomas & McCurtis, 2022).

9.8 THE HISTORICAL ROLE OF WOMEN IN AI

AI has evolved from a speculative idea into a powerful technology that is shaping nearly every sector of modern life. From business operations and healthcare diagnostics to education systems and public governance, AI is redefining how decisions are made, how services are delivered, and how economies function (Russell & Norvig, 2020). As this digital transformation accelerates, it brings with it both exciting possibilities and serious questions—especially about equity and gender.

Historically, women have always contributed significantly to the fabric of society—through caregiving, teaching, community-building, and workforce participation. Yet, despite their indispensable roles, women have faced persistent structural challenges: gender-based discrimination, wage inequality, limited access to leadership, and systemic bias in technology fields (UN Women, 2020). With AI now emerging as a dominant force in the digital economy, concerns are growing that without intentional design and policy safeguards, AI could replicate—or even amplify—existing gender inequalities (Noble, 2018; O'Neil, 2016).

However, it is equally important to recognize that women have not been strangers to the world of computing and AI. In fact, they were among its earliest pioneers. Ada Lovelace, a 19th-century mathematician, is widely regarded as the world's first computer programmer due to her work on Charles Babbage's Analytical Engine (Toole, 1998). During World War II, women like the ENIAC programmers—Kathleen McNulty, Jean Jennings Bartik, and others—performed essential programming work that laid the groundwork for modern computing (Abbate, 2012). Grace Hopper's contributions to the development of machine-independent programming languages fundamentally transformed how we interact with computers (Beyer, 2009). Despite these trailblazing achievements, women's roles in the development of AI have often been minimized, overshadowed by the dominance of male narratives in tech history.

Today, women remain underrepresented in AI-related fields. According to UNESCO (2020), only about 22% of AI professionals globally are women. Such imbalance is not caused by weaknesses or uninterestedness but rather a combination of various forms of systemic factors, such as inequitable gender stereotypes in early schooling, access to STEM shortages, discriminatory jobs and talent-management resources, and organizational cultures that frequently equate women in marginal positions (West et al., 2019). Unless these disparities are actively resolved, the development and deployment of AI is at risk of being flawed with only a limited number of experiences and views in mind.

However, to the women, AI remarkably has liberating prospects as well. The AI-based digital technologies are already opening new opportunities in remote work, online education, and entrepreneurial activity, especially for women in rural and underserved areas. Smart systems can enhance the availability of healthcare, finances, and law in the areas that women always experienced obstacles. As an example, microfinancing with the assistance of AI is providing women business owners access to credit in areas they were underserved by conventional banks (GSMA, 2021). On the same note, virtual learning environment and AI chatbots are bringing professional improvement closer to women who can combine the demands of family and care-taking.

In order to see to it that this potential is made into a reality, what is critical is that AI systems should be developed in such a way that in the process they end up providing potential users with: that is, inclusivity and equity by nature. Here, this implies the incorporation of gender sensitivity in data including the steps of the process of collection, algorithm design, testing, and deployment (Buolamwini & Gebru, 2018). It also demands robust policy frameworks to safeguard against any bias of algorithms and promote diversity in research and development teams in terms of gender (European Commission, 2020).

Further, it should be noted that collaboration between governments, academic institutions, and technology companies is required to embrace strategies that would proactively work in narrowing down the gender gap in AI. These can involve initiatives such as scholarship and mentorship support to girls in the sphere of STEM, auditing of the gendered character of AI systems, diversity-centric hiring practices, and increased representation of women in tech leadership positions (McKinsey & Company, 2020).

9.9 ENABLING WOMEN IN THE WORKPLACE THROUGH AI

AI is currently undergoing a reconfiguration of workplace dynamics, and it is influencing them in a fundamental manner. It holds the promise to make a huge difference in terms of helping women in the labor force to play a leadership role. As automation takes over routine, repetitive, physical endurance jobs. concentration of human work on strategic, inquiry and creative areas—most of which are becoming more gender-neutral, or gender-inclusive (ILO, 2021). This change can become very effective in enabling women to relocate to more higher-value and decision-making positions especially in fields that were long dominated by males, e.g., engineering, data science, and information and technology (United Nations, 2022; McKinsey Global Institute, 2019).

Remote and flexible work is one of the most revolutions AI has made. Women can also work on the tasks of professional nature remotely or in hybrid environments thanks to the use of AI-enabled tools, including smart scheduling platforms, virtual assistants, collaborative software, and automated workflow systems (World Economic Forum, 2021). In contexts in which women are still more than men to perform the role of carers, this tolerance may make it much easier to avoid the career cost that usually comes with family responsibilities (OECD, 2020). Surveys indicate that under permit to remote work, women are more likely to stay in the labor market, as well as, be willing to invest in leadership development and the ability to grow their careers in the long run (Accenture, 2020; UNDP, 2022).

Besides flexibility, AI-driven systems are improving personalized learning, mentorship, and career growth opportunities access. Tailor-made upskilling, Strengths based job matching, and partnering with industry mentors- offered through career building tools like Coursera, LinkedIn learning, or AI-powered HR platforms like Eightfold.ai which is proved to be a game changer in bringing women to the same level playing field (WEF, 2022). The platforms can identify and prevent any unconscious discrimination in recruitment and performance assessment, which means that female professionals receive fair assessments and have equal opportunities to promotion and acknowledgement (Binns et al., 2018; IBM, 2020).

Additionally, modern organizations applying AI-powered analytics are implementing gender equity measures to track progress on the specific metrics, thus assisting HR teams to analyze pay disparities, leadership diversity, promotions, etc. The information-based transparency will promote more women-friendly policies and practices that promote the success of women (Harvard Business Review, 2021). The adoption of these tools not only helps companies to create a more equal environment but also exposes them to the potential of multiple perspectives, which is essential in terms of innovation and performance.

The truth is that AI is not merely a technology or even revolution, but rather, it is a socio-economic change agent that can and will potentially comprise a shift in gender relations at the workplace. AI can be a great supporter of the global women empowerment and equality at work, provided it is used responsibly and inclusively.

9.10 WOMEN'S HISTORICAL CONTRIBUTIONS TO COMPUTING AND AI

Although AI is often associated with male-dominated tech industries, women have played a pivotal role in its development since the early days of computing. Some of the most groundbreaking contributions in computer science and AI come from women who laid the foundation for modern technology.

Pioneers in Computing

1. Ada Lovelace (1815–1852): Often considered the first computer programmer, Lovelace conceptualized algorithms for Charles Babbage's Analytical Engine. Her visionary work laid the foundation for modern programming (Fuegi & Francis, 2003).
2. Grace Hopper (1906–1992): A pioneer in software development, Hopper invented one of the first compilers, which eventually led to the development of programming languages like COBOL. Her work revolutionized the way computers were programmed (Beyer, 2009).
3. Margaret Hamilton (b. 1936): Hamilton led the software engineering team responsible for NASA's Apollo program, developing the onboard guidance system that enabled the moon landing. She is credited with coining the term "software engineering," highlighting the discipline's importance in technology development.
4. Jean Sammet (1928–2017): A key figure in programming language development, Sammet was instrumental in the creation of FORMAC, an early programming language used in symbolic mathematical computations.

Modern-Day Female Leaders in AI:

- Dr. Fei-Fei Li: Dr. Li is considered one of the best researchers in computer vision, who created ImageNet, a dataset so big that it revolutionized the use of deep learning. These works have made major contributions to improving the capacity of AI to analyze visual data and achieve progress in facial recognition and autonomous cars, as well as medical imaging (Li & Li, 2020).
- Mira Murati: Murati serves as the Chief Technology Officer of the OpenAI company and has been instrumental in the creation of ChatGPT and DALL·E AI models. Her work in AI research made the advancements of generative AI a mainstream reality, including powerful natural language processing and image creation.

They underscore how women have made their most vital contributions to the AI and computing field, though they are usually not given due recognition and thus underrepresented. By acknowledging their success, we will be able to encourage the young girls in the next generations to be interested in AI, machine learning, and technology-related careers.

Women were involved on the ground in the historic achievements of AI and computing fields, and despite this, they still struggle in this domain. Since AI has already started playing a bigger role in our day-to-day activities, it is necessary to make sure that gender equality is one of the priorities when developing and using AI.

Through the use of AI in the construction of more inclusive working environments, the empowerment of women in management positions, and the acknowledgment of women who pioneered AI in history, the society will be on its path to becoming more diverse and fairer in the future. The inclusion of woman in AI is not only about fairness, it is also unarguably essential to make AI systems ethical, bias-free, and reflective of their whole population. We can create conditions in which women will excel in AI and shape its future by means of education, mentorship, and the development of such policies.

9.11 CASE STUDY: INFLUENTIAL WOMEN IN AI—PIONEERING CONTRIBUTIONS AND IMPACT

Historically, AI has been governed by a male predominance, but since the past few years, there have been some female powerhouses dominating the field who have defined the field and pioneered the usage of AI technologies. The case study revolves around two personalities that include Dr. Fei-Fei Li, who pioneered computer vision and deep learning, and Mira Murati, the Chief Technology Officer of OpenAI, whose contribution in this field is instrumental in the creation of powerful AI models. They bring together the great importance of women in AI that is generally unrepresented.

9.11.1 Case 1: Dr. Fei-Fei Li—Advancing Computer Vision and Deep Learning

Dr. Fei-Fei Li is a reputable computer scientist who has proved to be a pioneer in the development of machine learning and computer vision. She is a cofounder of the ImageNet, a massive dataset that revolutionized AI research by enabling the deep learning models to achieve impressive accuracy in image recognition (Li & Li, 2020). The development of the ImageNet gave way to the deep learning revolution, which will enhance the AI capacities in sectors like healthcare, facial recognition, and driverless cars.

Dr. Li has also been a big proponent of ethics in development of AI and inclusion of women in technology. She was one of the co-founders of the nonprofit organization AI4, geared at improving diversity and inclusion in the AI field, through mentoring and educational opportunities to underrepresented populations, which includes women (AI4, 2021). Her work has been instrumental in spreading the proactive adoption of AI and encouraging more women to go into AI as a career field.

9.11.2 Case 2: Mira Murati—Leading AI Innovation at OpenAI

Mira Murati is the Chief Technology Officer (CTO) of one of the significant AI innovators at OpenAI. She has created models of AI such as ChatGPT. This is the case with the GPT-4 (and DALL·E), which have revolutionized natural language processing (NLP) (OpenAI, 2023). During her tenure, OpenAI has been the first of its kind to work on advancements in the field of AI-based conversation, AI-based art content-creating, and multimodal AI.

In addition to technological innovation, Murati has made note of the ethical and responsible use of AI as well. She has been the key figure in crafting OpenAI policies on AI safety and ensuring that the AI is not used in a fashion that will harm societal development, regardless of risks to misinformation and biasness (Murati, 2023).

9.12 IMPACT AND CHALLENGES

Dr. Fei-Fei Li and Mira Murati have both exemplified the vitality of women in the AI field even though the problem of gender disparity is still a sore issue concerning the extent of women in AI. The inclusion of women in the leadership of AI and technical positions remains insufficient, and this situation requires women-friendly policies, mentorship, and the ability to make female charismas visible. Their text highlights the significance of the development of a fair AI industry in which women get the same chances to impact the technological future.

The works by Dr. Fei-Fei Li and Mira Murati can be considered regarding female revolutionary contributions to the field of AI. They have left a mark in the industry in terms of their breakthroughs in computer vision, deep learning, and AI model development. It is important to notice and encourage the contribution of women to AI as a method of advancing innovations and ethical uses, as well as diversification of the future of technology.

9.13 CHALLENGES FACED BY WOMEN IN AI AND STEM

These people make important contributions to the field of science, technology, engineering, and mathematics despite some of their drawbacks. There remains the issue of women underrepresentation in AI. The World Economic Forum (2021) believes that one-half of all AI professionals worldwide lack the in-demand skills of women. This gender divide is propelled by some of the major issues that affect the position of women in this discipline.

1. Educational Barriers and Gender Stereotypes
   The traditional stereotypes support the opinion that many young girls are discouraged to pursue STEM careers. At an early age, they can be informed that science and technology are male-dominated subjects that will cut down their confidence and interest in subjects such as mathematics and computer science. Also, when there are no female AI and technology role models, girls would not have role models that would be in their fields of profession. Consequently, there are fewer women enrolling in higher education and participating in AI-affiliated areas of work (UNESCO, 2020).
2. Workplace Disparities and Career Growth Challenges
   When women have entered the field of AI and STEM, there is a lot of inequality in the workplace that they end up getting. Gender discrimination in the field of employment may complicate the process of giving women access to working in areas related to AI, and one of the remaining problems is pay inequality. McKinsey & Company (2020) also states that women are paid a lot less than men with the same education and experience in the field of AI and technology. Moreover, a number of women cannot get promotions, and this fact does not allow them to gain the top position of a leader.

3. Underrepresentation in Leadership and Decision-Making
   Only a small percentage of women hold positions of power in AI companies, so decision-making and innovation in the field are also affected. AI policies and developments might not focus on gender aspects when the leadership teams are dominated by the male gender. This is also due to the fact that there is a shortage of women occupying leadership positions which will also lead to a shortage of role models and mentors who will discourage young women from going into the industry, leading to the same cycle of maintaining the gender gap (Beyer, 2009).
4. Algorithmic Bias and Gender Inequality in AI Systems
   The bias in the society is likely to be promoted when training AI models with biased datasets. Consequently, the AI-powered program can support the phenomena of gender stereotyping during human resources recruitment, credit provision, health prescriptions, and all other automated decisions. As an example, research results have ranked the AI hiring tools simply because they tend to favor hiring the candidates of the male gender simply because the old hiring statistics were still biased in the past (Bolukbasi et al., 2016). This problem demonstrates why there should be an immediate change to diverse AI teams that can recognize and eliminate bias in machine learning models.

Even though women have become more represented in AI and STEM, there is still a problem with their opportunity to participate and progress. Better policies should be implemented to enhance gender equality, and women in the field of AI should be mentored, and more people should be educated about the biases of AI systems so that the challenges can be overcome. We can make the AI industry more inclusive and diverse, and because of this, everyone will benefit and will be able to seek gains in the future technological progress, not being affected by gender issues.

9.14 CONSEQUENCES OF GENDER IMBALANCE IN AI

Gender diversity in the AI sector has far-fetched implications, especially on the way AI systems make decisions that impact the lives of people. Absence of varied opinions in the worlds of AI development means that unfairness and discrimination may follow due to lack of perspectives that formed the algorithm and data. Such inequity is particularly problematic in facial recognition, hiring, and language modeling programs, where AI might solidify inequity in the entire society, but not eradicate it.

1. Bias in Facial Recognition Technology
   Facial recognition software bias is one of the topics that is most worrying in relation to AI. The studies conducted have revealed that such systems are likely to accurately apply to white men but fail miserably when it comes to women with darker complexion. The error rate of facial recognition technology with darker-skinned women is up to 34%, whereas it practically does not mistake white men (Buolamwini & Gebru, 2018).

Such a prejudice has disastrous real-life implications. AI identification errors in a security system can therefore lead to one being detained or not getting access to valuable services. Equally, try as they may, companies that hire people based on facial recognition may end up discriminating against women and minorities, thus limiting their chances of employment.

2. Gender Bias in AI-Driven Recruitment Tools
   The use of AI hiring systems has become popular among companies to automate recruitment processes and help reduce both discriminatory and unwanted hiring choices; however, these systems can show gender biases based on how they were used in the past. One of the most famous ones is the Amazon AI hiring tool that was found to have a bias against women in hiring. The AI system was modeled using the history of the hiring data and, in the past, male applicants were always favored in technical positions. Consequently, the system penalized resumes that carried the component of the words indicating women intrusion, e.g., women chess club or women colleges (Dastin, 2018).

   This form of discrimination in hiring instruments reduces the possibilities of women pursuing careers and perpetuates the inequality that exists in the workplace. Without stopping it, the biased AI hiring system may render the process of women entering fields of males even more difficult, such as technology, engineering, or AI itself.
3. Reinforcing Gender Stereotypes in Language Models
   Language models in AI are trained based on enormous quantities of textual information on the Internet, like the model used in chat scripts and virtual assistants. Unluckily a significant part of this information is stereotypical relating to all the gender types and it is enhanced and reproduced by the models of AI. It has been established that AI systems tend to connect males with leadership, ambition, and intelligence and females with caregiving and emotional support and domesticity (Bolukbasi et al., 2016).

As an example, AI models may be biased toward male candidates when assigning job descriptions or offering to take leadership positions since, in the past, the results show male leaders. On the same note, the textual output of AI could even use terms like relabeling nurses as women and engineers as men, contributing to the old stereotyping.

The fact that people define it with a gender imbalance directly affects the way AI systems operate, giving biased and discriminatory results. To make AI a tool of justice and improvement, there must be an attempt to eradicate the biases present in facial recognition software, employment platforms, and language-processing software. This is what is needed of various teams working on the creation of AI, better training data, and constant efforts dedicated to the process of finding and removing biases in the AI system. By making AI gender diverse we will be able to develop more equitable, open technologies, and use them to benefit the entire society.

9.15 STRATEGIES FOR ENHANCING WOMEN'S PARTICIPATION IN AI

There is a need to set special steps in each direction to get rid of the gender gap in AI: education, professional development, and entrepreneurship. We will be able to build a more inclusive and diverse AI industry by taking measures that will promote the involvement of women in it. Some of the most prominent ways of attaining this objective are as follows:

1. Encouraging STEM Education for Girls
   It is crucial to teach AI, coding, and related topics to young girls early in their studies, thus boosting their confidence towards technology. The reason why there are not many girls who want to have a STEM career is that all the negativity surrounding the stereotypes about the field and the lack of visible role models discourage them from doing so. Mentorship programs like Girls Who Code and AI4 give practical experience and access to the utilization of this technology in practice. These programs will continue to eliminate the disparities and motivate more young girls to consider the world of technology (Resnick, 2019).

   Schools can also help by adding AI and coding classes to their classes, and girls should have the same access to this learning process. Such additional practices as conducting workshops, hackathons, and competitions in AI conducted exclusively among young women can also help develop their interest in the sphere and their competence.
2. Expanding Mentorship and Networking Opportunities
   Access to good mentors and professional networks is necessary because women have to be able to move on in their career in the area of AI. Mentorship Programs: The mentorship program has enabled women to interact with leaders in the industry and seek some guidance and career advice and also assist in honing their skills. The organizations such as Women Techmakers designated by Google and AnitaB.org offer guidance and leadership training as well as events to strive to help women in AI and technology.

   Such network possibilities enable women to create worthwhile professional connections, learn with other participants, and be noticed within the field. Events such as the Grace Hopper Celebration, the largest conference of women in computing, are also where women are able to get a voice to share their ideas, to find employment, and to be awarded for their contributions to the field.
3. Boosting Investment in Women-Led AI Startups
   This is the key to propelling innovation and generating economic activity in the form of female entrepreneurs developing AI. In the past, there has been an under-proportionate amount of venture capital invested in startups headed by women. With more investment in businesses by women, groundbreaking innovations can be made in fields like healthcare, education, and finance.

One such example is the creation of the FemTech industry, which is specifically addressed with AI-based solutions to women's health-related problems. The latest technology in the area of reproductive health, motherhood, and customized medicine demonstrates the ability of female-led technological companies to transform an industry. The lack of funding, incubators, and accelerator programs is a severe issue that affects women. The gap in finance can be closed and further intake of female entrepreneurs into AI ventures is achievable through preparing them with relevant access (Forbes, 2022).

4. Promoting Inclusive Policies in Workplaces

   Creating a more equitable AI industry requires companies to implement workplace policies that support gender diversity and inclusion. Several key areas should be addressed:

   - Diversity-Driven Hiring Practices: Employers should adopt unbiased recruitment policies that ensure fair hiring and promotion opportunities for women. Implementing blind recruitment—where resumes are evaluated without names or gender indicators—can help eliminate bias in the hiring process (McKinsey & Company, 2020).
   - No Gender-Based Limits on Access to Investment and Leadership: Women should have equal opportunities to access investments, leadership, and career development just like men. Judicious efforts should be made by financial institutions and venture capital firms to ensure that there is increment in funding of women-led AI startups so that these companies can grow and scale their ventures (Forbes, 2022).
   - Designing Bias-Free AI: AI systems must be assembled fairly. The optimized aspect of the research should be to reduce gender biases in the AI and train the models to be egalitarian and diverse. The organizations should make periodic audits on their AI systems that may have any embedded bias in a way that would increase discrimination (Bolukbasi et al., 2016).

It is essential to increase the involvement of women in AI not only because it is morally right but also to innovate and influence the ethical design of AI. We can improve the situation by motivating girls to pursue STEM in their education, promote mentorship and connections, encourage entrepreneurship among women, and introduce non-discriminatory policies in the workplace that can lead to a future where female participation in developing AI will be an equal and significant force. Diversity among the workforce in AI is one way of attaining good ownership, ethical, and inclusive AI systems, thereby advancing society as a whole.

9.16 CASE STUDIES OF SUCCESSFUL GENDER INCLUSION INITIATIVES IN AI

There have been a number of international initiatives that have started to narrow the gender gap in AI and technology. The target of these programs would be to reduce the gender gap especially by offering education, mentorship, networking, and career growth opportunities. These are some of the impressive case studies of initiatives that have endeavored to advance gender inclusion in AI.

1. Girls Who Code
   Girls Who Code is a nonprofit organization based all over the world, founded and run to close the gap in women working in technology by teaching the basics of coding to young girls. Reshma Saujani started the organization in 2012, and it provides after-school programs, summer coding camps, and full-time mentoring programs to girls with different backgrounds.

   Impact:
   Over thousands of girls have received assistance in cultivating technical skills through the program, with a subsequent rise in their confidence and level of interest in careers in the AI, software development, as well as other STEM areas. The study indicates that Girls Who Code alumni have a higher chance of attaining a degree in computer science than in the rest of the nation (Resnick, 2019). This initiative has included women more in the AI industry by giving free and easy-to-access resources that the initiative now provides so that young women can know coding better and feel more empowered to go into the industry.
2. AI4
   IAI organizes IAIEAI4ALL, a non-profit which addresses the problem of inequality in education and aims to educate members of underrepresented groups about AI, women, minorities, and individuals of low income. The program works in collaboration with universities and industry leaders in order to offer training, research opportunities as well as career mentorships regarding AI issues in form of practical projects

   Impact:
   AI4 helps to change this system by establishing diversity in education where underrepresented groups can build an inclusive AI ecosystem and have a voice in the innovation. The participants also get exposure to AI concepts, ethics, and real-life applications, and this puts them in line in the careers involving AI research and development. UNESCO (2020) explains that AI4 has managed to bring more diversity in terms of AI-related fields because it seeks to empower young women to enter the engineering workforce in machine learning, data science, and computer vision.
3. Women in Machine Learning (WiML)
   WiML is a global organization that assists in recruiting women in the area of AI and machine learning. It was established to offset the deficiency of females in AI research, by providing networking possibilities, mentorship services, and financial support for conference visits.

   Impact:
   WiML organizes workshops within the biggest AI conferences, such as the Neural Information Processing Systems (NeurIPS) and the International Conference on Machine Learning (ICML), which allow women to have a

platform on which they can present their work and where they can form connections and have visibility. The work done has led to the appearance of a growing network of women AI experts to improve their attendance in higher education and the industry.

4. Google's Women Techmakers
   Women Techmakers is an initiative launched by Google to make women successful in the technological sector through provision of training and leadership opportunities, and links women to networks. The program is also providing grants, scholarships, and mentorships that will attract more women into careers such as AI, software engineering, and other job opportunities revolving around tech.

   Impact:
   Women Techmakers has built a community of female professionals in the AI sphere worldwide, allowing them to network with industry leaders and initiate new learning and find job opportunities. Women are also encouraged to talk about their problems, share experience, and promote the inclusion of women in the field of technology via annual summits and workshops organized by Google. As per Google (2022), the initiative has led to more women spearheading the sphere of AI, as it helped them in improving their careers.
5. UNESCO's STEM Gender Equality Program
   The STEM Gender Equality Program by UNESCO is an international initiative focused on executing strategies to promote the interests of women regarding their inclusion in STEM fields of study and their practice. The program collaborates with governments, schools, and other private organizations to help them establish inclusive learning cultures, establishing gender-sensitive teaching resources and managing any biases in the field of STEM education.

   Impact:
   UNESCO has used advocacy and policy suggestions to ensure that the STEM curriculum in all countries has equal opportunities to groups of girls to take science and technology courses and to understand how teachers should be trained on the contents of STEM. According to UNESCO (2020), this program has served to advance the enrolment of women in STEM classes especially in countries that have traditionally had minimal female contribution in AI and technologies.

Such programs prove that it is possible to bring a positive change through specific programs to teach, guide, and empower women in the field of AI. Such initiatives as Girls Who Code, AI4, WiML, and Women Techmakers and the STEM Gender Equality Program of UNESCO have been instrumental in enabling the removal of obstacles and the establishment of an AI industry that is much more diverse and inclusive. Investing into such initiatives further will help to make sure that women will be equally offered opportunities and thus will also play a major role in the future of AI.

9.17 CONCLUSION

The end of the chapter highlights the eminent importance that women have made in the exposure and enrichment of AI amidst the ongoing gender-based unfairness and systematic difficulties. The past roles of women in AI have also been pillars, yet the past inequities, institutional impediments, as well as social conventions have normally understated their roles in the field as well as their contribution to the same.

The chapter presents the argument that representation of women in AI should not only be a question of fairness but also the main source of innovation and diversity. AI systems are applied almost to every sphere of our lives including healthcare, education, finance, and even the sphere of government. Nevertheless, in case such systems are created by diverse groups working in stereotyped environments, there exists the emergence of gender biasness and escalation of the current disparity. Women being locked out of AI development has raised the potential of ISO bad algorithms which can include prejudiced face scanners, unfair recruitment systems, and gender-biased communication programs. The importance of diversity in AI can be presented on the basis of these questions regarding the necessity of ethical, inclusive, and successful technological development.

The chapter also commemorates the accomplishments of groundbreaking women in AI like Dr. Fei-Fei Li and Mira Murati who have contributed a lot in the fields of computer vision, deep learning, and AI studies. Their achievements are examples of inspiration to the new generations, and they underline the necessity of women leadership in the future of AI.

The chapter concludes, in a positive manner, by imagining a world where a woman and her place are taken care of in AI. Diversifying society by breaking down the obstacles and establishing an inclusive world is the way to create an AI system that will be conditioned to acknowledge the diverse views, reduce the risks of biases, and produce a fair technological development that will benefit everyone. A solution to the gender disparity in AI is not only a social responsibility, but this concern is necessary to achieve ethical, responsible, and successful development of AI.

REFERENCES

Abbate, J. (2012). *Recoding Gender: Women's Changing Participation in Computing*. MIT Press.

Accenture. (2020). *The Future of Work: A Journey to 2022.*

Baker, P. (1981). The Equal Pay Act of 1963: A decade of enforcement. *Industrial and Labor Relations Review*, 34(3), 356–370.

Berg, Florian, Kölbel, Julian and Rigobon, Roberto. (August 15, 2019). Aggregate Confusion: The Divergence of ESG Ratings Forthcoming Review of Finance, Available at SSRN: https://ssrn.com/abstract=3438533 or http://dx.doi.org/10.2139/ssrn.3438533

Beyer, K. W. (2009). *Grace Hopper and the Invention of the Information Age*. MIT Press.

Bianchi, S. M., Milkie, M. A., Sayer, L. C., & Robinson, J. P. (2000). Is anyone doing the housework? Trends in the gender division of household labor. *Social Forces*, 79(1), 191–228.

Binns, R., Veale, M., Van Kleek, M., & Shadbolt, N. (2018). Fairness in algorithmic decision making. *Proceedings of the 2018 CHI Conference on Human Factors in Computing Systems (CHI'18)* doi:10.1145/3173574.3174014, ISBN: 978-1-4503-5620-6. 14 Pages.

Bolukbasi, T., Chang, K. W., Zou, J. Y., Saligrama, V., & Kalai, A. (2016). Man is to computer programmer as woman is to homemaker? Debiasing word embeddings. *30th Conference on Neural Information Processing Systems (NIPS 2016), Barcelona, Spain.*

Buolamwini, J., & Gebru, T. (2018). Gender shades: Intersectional accuracy disparities in commercial gender classification. *Proceedings of the 1st Conference on Fairness, Accountability and Transparency*, PMLR 81, 77–91.

Crenshaw, K. (1991). Mapping the margins: Intersectionality, identity politics, and violence against women of color. *Stanford Law Review*, 43(6), 1241–1299.

Dastin, J. (2018). *Amazon Scraps Secret AI Recruiting Tool that Showed Bias Against Women.* Reuters.

Eagly, A. H., & Carli, L. L. (2007). *Through the Labyrinth: The Truth About How Women Become Leaders.* Harvard Business Review Press.

Eagly, A. H., & Karau, S. J. (2002). Role congruity theory of prejudice toward female leaders. *Psychological Review*, 109(3), 573–598.

Eagly, A. H., Johannesen-Schmidt, M. C., & van Engen, M. L. (2003). Transformational, transactional, and laissez-faire leadership styles: A meta-analysis comparing women and men. *Psychological Bulletin*, 129(4), 569–591. https://doi.org/10.1037/0033-2909.129.4.569

European Commission. (2020). *White Paper on Artificial Intelligence – A European Approach to Excellence and Trust.*

Flexner, E., & Fitzpatrick, E. (1996). *Century of Struggle: The Woman's Rights Movement in the United States.* Harvard University Press.

Forbes. (2022). *The Rise of FemTech: Women-Led Innovation in AI and Health Technology.*

Forbes, G. (1996). *Women in Modern India.* Cambridge University Press.

Freeman, J. (1975). *The Politics of Women's Liberation: A Case Study of an Emerging Social Movement and Its Relation to the Policy Process.* David McKay Company.

Friedan, B. (1963). *The Feminine Mystique.* W. W. Norton & Company.

Fuegi, J., & Francis, J. (2003). Lovelace & Babbage and the creation of the 1843 "Notes". *IEEE Annals of the History of Computing*, 25(4), 16–26.

Garikipati, S., & Kambhampati, U. (2020). Leading the Fight Against the Pandemic: Does Gender Really Matter? *World Development*, 105, 104708.

Google. (2022). *Women Techmakers: Empowering Women in Technology.*

GSMA. (2021). *Connected Women: The Mobile Gender Gap Report.*

Harvard Business Review. (2021). *Why Gender Equity Is More Than a "Women's Issue".*

Hewlett, S. A. (2002). *Creating a Life: Professional Women and the Quest for Children.* Talk Miramax Books.

Hewlett, S. A., Marshall, M., & Sherbin, L. (2010). *The Sponsor Effect: Breaking Through the Last Glass Ceiling.* Harvard Business Review Research Report.

Hochschild, A. R., & Machung, A. (1989). *The Second Shift.* Viking Penguin.

Ibarra, H., Carter, N. M., & Silva, C. (2010). Why men still get more promotions than women. *Harvard Business Review*, 88(9), 80–85.

IBM. (2020). *AI Fairness 360 Open Source Toolkit.*

ILO. (2021). *Gender Equality and the Future of Work.*

Inter-Parliamentary Union (2023). https://www.ipu.org/resources/publications/about-ipu/2024-03/impact-report-2023

Kanter, R. M. (1977). *Men and Women of the Corporation.* Basic Books.

Lerner, G. (1993). *The Creation of Feminist Consciousness.* Oxford University Press.

Li, F. F., & Li, J. (2020). *Visual Intelligence: How AI Sees the World.* MIT Press.

McKinsey & Company. (2020). *Diversity Wins: How Inclusion Matters.*

McKinsey Global Institute. (2019). *The Future of Women at Work: Transitions in the Age of Automation.*

Murati, M. (2023). *AI Ethics and Responsibility: OpenAI's Approach.* OpenAI Research Blog.

Noble, S. U. (2018). *Algorithms of Oppression: How Search Engines Reinforce Racism.* NYU Press.

Northouse, P. G. (2021). *Leadership: Theory and Practice* Ninth Edition. SAGE Publications, Inc.

Nussbaum, M. (2000). *Women and Human Development: The Capabilities Approach.* Cambridge University Press.

O'Neil, C. (2016). *Weapons of Math Destruction: How Big Data Increases Inequality and Threatens Democracy.* Crown Publishing.

OECD. (2020). *Women at the Core of the Fight Against COVID-19 Crisis*

OpenAI. (2023). *Advancements in Generative AI: The Role of ChatGPT and Beyond.*

Purvis, J. (1995). *Women's History: Britain, 1850–1945.* UCL Press.

Resnick, M. (2019). The Next Generation of Scratch Teaches More Than Coding. *EdSurge*, January 2019.

Ridgeway, C. L. (2001). Gender, status, and leadership. *Journal of Social Issues*, 57(4), 637–655.

Russell, S., & Norvig, P. (2020). *Artificial Intelligence: A Modern Approach* (4th ed.). Pearson.

Ryan, M. K., & Haslam, S. A. (2005). The Glass Cliff: Evidence that Women are Over-Represented in Precarious Leadership Positions. *British Journal of Management, 16*(2), 81–90. https://doi.org/10.1111/j.1467-8551.2005.00433.x

Scott, J. W. (1996). *Only Paradoxes to Offer: French Feminists and the Rights of Man.* Harvard University Press.

Thomas, D. A., & McCurtis, D. M. (2022). *An Intersectional Approach to Advancing Equity in Tech Leadership.* MIT Sloan Management Review.

Toole, B. A. (1998). *Ada, the Enchantress of Numbers.* Strawberry Press.

UN Women. (2020). *Gender Equality: Women's Rights in Review 25 Years After Beijing.*

UNDP. (2022). *Gender Equality and Inclusive Digital Economies*

UNESCO. (2020). *Artificial Intelligence and Gender Equality: Key Findings of UNESCO's Global Dialogue.*

United Nations. (2022). *Women and the Future of AI: Challenges and Opportunities.*

West, S. M., Kraut, R., & Chew, H. E. (2019). *Discriminating Systems: Gender, Race and Power in AI.* AI Now Institute.

World Economic Forum. (2021). *Global Gender Gap Report 2021.*

World Economic Forum. (2022). *Upskilling for Shared Prosperity: A Vision for Closing the Skills Gap.*

World Economic Forum's Global Gender Gap Report (2023). https://www.weforum.org/publications/global-gender-gap-report-2023/

Index

Note: **Bold** page numbers refer to tables.

S

T

U

V

W